GRAND CAYMAN

THREATENED PLANTS OF THE

CAYMAN
ISLANDS
THE RED LIST

THREATENED PLANTS OF THE
CAYMAN ISLANDS
THE RED LIST

Frederic J. Burton

Kew Publishing
Royal Botanic Gardens, Kew

PLANTS PEOPLE
POSSIBILITIES

First published in 2008
by Royal Botanic Gardens, Kew
Richmond, Surrey, TW9 3AB, UK
in association with the Cayman Islands Department of the Environment

www.kew.org

ISBN 978-1-84246-220-1

British Library Cataloguing in Publication Data
A catalogue record for this book is available from the British Library.

PRODUCTION EDITORS: Michelle Payne and Sharon Whitehead

DESIGN, TYPESETTING AND PAGE LAYOUT: Jill Bryan

PHOTOGRAPHY CREDITS Front cover: *Agave caymananensis* flowers (photograph by Frederic J. Burton). Back cover left to right: *Turnera triglandulosa* flower (photograph by Dr. Mat DaCosta-Cottam), *Tabernaemontana laurifolia* fruit and *Cordia sebestena* flower (photographs by Frederic J. Burton). Endpapers: Aerial photographs of Grand Cayman (front) and the Sister Isles (back) prepared by Jeremy Olynik. Half title: Silver Thatch (photograph by Patrick Broderick). Main title: *Myrmecophila thomsoniana* var. *thomsoniana* (photograph by Frederic J. Burton).

For information or to purchase all Kew titles please visit
www.kewbooks.com or email publishing@kew.org

All proceeds go to support Kew's work in saving the world's plants for life

Mixed Sources
Product group from well-managed forests and other controlled sources
www.fsc.org Cert no. CQ-COC-000012
© 1996 Forest Stewardship Council

The paper used in this book contains material sourced from responsibly managed and sustainable commercial forests, certified in accordance with the FSC (Forestry Stewardship Council), and manufactured under strict environmental systems, the international ISO 14001 standard, EMAS (Eco-Management and Audit Scheme) and the IPPC (Integrated Pollution Prevention and Control) regulation.

Printed and bound in Italy by Printer Trento

CONTENTS

ACKNOWLEDGEMENTS   6

FOREWORD   7

SUMMARY
  Table 1: Conservation status summary for the Cayman Islands'
  native flora   8

1. INTRODUCTION
  The Cayman Islands   10
  Vegetation   11
  Classification of plant communities   13
  Forests   13
  Shrublands   16
  Dwarf-shrublands   17
  Herbaceous   17
  Deforestation and endangerment   20
  Alien invasion   21
  Protected areas   22
  Table 2: Percentage of the land area under environmental
  protection   28
  Notes on methods   29
  Key to the plant description pages   30

2. PLANT DESCRIPTIONS
  The endemic flora   32
  Threatened near-endemic flora   61

3. RED LIST TABLES
  for the Cayman Islands native flora   77

BIBLIOGRAPHY   102

INDEX OF SCIENTIFIC NAMES   103

INDEX OF COMMON NAMES   105

ACKNOWLEDGEMENTS

Data used in these Red List assessments depended in part on outputs from WWF-UK Project No. 97068 (Biodiversity Survey of the Cayman Islands, National Trust for the Cayman Islands). Kevin Eden, Penny Clifford, Gerald F. Guala III, David Healy, Richard Burton, Michael Bradley Jnr, John Haines, and Brian Davies assisted the author in fieldwork for that survey.

Ann Stafford, Frank Roulstone III, Wallace Platts, Carla Reid, and Lois Blumenthal contributed valuable additional field observations on specific taxa.

Dr. Mat DaCosta-Cottam (Cayman Islands Department of Environment), and Dr. Colin Clubbe (Royal Botanic Gardens, Kew), provided advice and support throughout this work. The maps were prepared by Jeremy Olynik. These include aerial photography and data layers reproduced by kind permission of the Lands & Survey Department, Cayman Islands Government.

This work was funded by a grant from the British Government's Overseas Territories Environment Programme to the Cayman Islands Department of Environment (OTEP: CAY-001, Red List assessment of Cayman Islands' native flora for legislation and conservation planning).

All photographs are by the author other than those listed here:
Gary White (www.infocusphotosltd.com); page 18 - Aerial view of Grand Cayman.
John F. Binns; pages 61 and 64 - *Cionosicyos pomiformis* fruit.
Dr. Mat DaCosta-Cottam; page 35 - *Pectis caymanensis* var. *caymanensis* habit, page 38 - *Hohenbergia caymanensis* habit, page 46 - *Salvia caymanensis* (both photographs), page 55 - *Banara caymanensis*, page 59 - *Turnera triglandulosa* (both photographs), page 68 - *Encyclia phoenicia* (both photographs).
Penny Clifford; page 47 - Watercolour of *Dendropemon caymanensis*.
Ann Stafford; Page 56 - *Casearia staffordiae* fruits.

FOREWORD

Plants are an essential component of the world's biological diversity and provide a unique set of vital resources for human well-being. Despite their multi-faceted roles and functions in almost every element of human activities, plants are subjected to severe threats globally. These threats include habitat loss and fragmentation, over-exploitation, invasion by non-native species (a particular threat on islands such as the Caymans), and the increasingly worrying threat from global climate change.

The Global Strategy for Plant Conservation was unanimously adopted at the sixth meeting of the Conference of the Parties to the Convention on Biological Diversity held in The Hague in April 2002. The specific goal of this strategy is to halt the current and continuing loss of plant diversity. Its 16 challenging targets provide a framework for action within an equally challenging timeframe – by 2010. The underpinning targets of the Strategy are: Target 1, a widely accessible working list of known plant species, as a step towards a complete world flora; and Target 2, a preliminary assessment of the conservation status of all known plant species, at national, regional and international levels. We need to know what plant species exist and what their status is in the wild in order to determine what their conservation needs are, and how to plan and manage the protection of plant diversity for future generations.

In the context of the UK's plant diversity, the Overseas Territories are real gems that have significantly greater species and habitat diversities than exhibited in mainland UK. The three islands comprising the Cayman Islands (Grand Cayman, Little Cayman and Cayman Brac) support 415 native taxa in a land area of a little over 260 km^2; 29 of these taxa are uniquely Caymanian.

RBG Kew has a long history of involvement with the Overseas Territories. The UK Overseas Territories Programme at Kew is a particularly active cross-departmental team who have been collaborating with partners in Overseas Territories for many years. We are very pleased to be publishing the Cayman Islands Red List – the first full Red List for a UK Overseas Territory, making the Caymans the first Territory to fully achieve Targets 1 and 2 of the Global Strategy for Plant Conservation. I am pleased to welcome this publication into the conservation literature and hope that it will be the first in a series from the UK Overseas Territories. It can provide a roadmap for the conservation of the unique plant diversity of the Cayman Islands and is directly applicable to other Territories and the wider plant conservation community globally.

Professor Stephen Hopper FLS
Director, Royal Botanic Gardens, Kew

SUMMARY

There are 415 taxa (species and varieties) of plant believed to be truly native to the Cayman Islands. These are the plants which formed the original, ancient flora of Grand Cayman, Little Cayman and Cayman Brac.

Over the two to three million years between the Cayman Islands' most recent emergence from the sea and the arrival of humans, 29 species and varieties of plant had evolved to local conditions to such a degree that they are now regarded as unique (endemic) to the Cayman Islands. If these plants are lost from the Cayman Islands, they will be lost from the world.

The majority of the Cayman Islands native plants are exquisitely adapted to the dry forests, shrublands, and wetlands in which they have evolved. Many are incapable of surviving naturally outside these complex natural communities.

Yet the Caymanian landscape is being subjected to large-scale deforestation. Human land uses ranging from housing and roads to golf courses and quarries are displacing natural forests, shrublands and wetlands at an accelerating rate. By 1998, Grand Cayman was already 37% deforested. Cayman Brac was close behind at 26% deforestation. Even Little Cayman, with its far smaller human population, had suffered 19% loss of its natural vegetation cover.

The deforestation of the Cayman Islands is ongoing, and appears to be accelerating. As a result, at least 46% of the Cayman Islands native plants are now threatened with extinction.

	CR	EN	VU	Total threatened	NT	LC	DD	All taxa
Endemic taxa	16	5	3	24	1	3	1	29
Near-endemics	6	6	3	15	0	3	6	24
Local regionals	7	9	12	28	0	5	11	44
Other taxa	55	43	27	125	5	120	68	318
Total	84	63	45	192	6	131	86	415
	20%	15%	11%	46%	1%	32%	21%	

Table 1: Conservation status summary for the Cayman Islands' native flora. See p. 31 for an explanation of Red List status abbreviations.

Threatened Plants of the Cayman Islands is a Red List assessment of the Cayman Islands' native flora. The book contains three main sections. The introductory chapter outlines the vegetation types found on the Islands, provides an overview of the risks to the flora, and lists the areas of high biodiversity that have been placed under environmental protection. Data assessment methods are explained in the notes on methods. A full explanation of the plant description pages is provided on pp. 30–31. Section two consists of the detailed, illustrated plant descriptions, covering the Cayman Islands' endemic flora and the threatened near-endemic flora. Finally, a Red List table covering the entire native flora is included on pp. 77–101.

Critically endangered Old George (*Hohenbergia caymanensis*, p. 38) at a deforestation margin. George Town, Grand Cayman

INTRODUCTION
The Cayman Islands

Perched on the southern rim of the North American tectonic plate, approximately 200 km south of Cuba, the three Cayman Islands are the emergent peaks of a partially submerged mountain range which extends from Cuba's Sierra Maestra, across the north-western Caribbean to the Misteriosa bank, and onwards into the Gulf of Honduras. Tiny by comparison to Cuba and neighbouring Jamaica, Grand Cayman is 197 km² in area. Little Cayman and Cayman Brac are 28.5 km² and 38 km², respectively.

Geologically, these are islands of limestone and dolostone rocks, derived from ancient seabed sediments. Over the history of their original uplift and subsequent changes in sea level, they have existed variously as hills rising steeply out of the sea, or at times as submerged shallows.

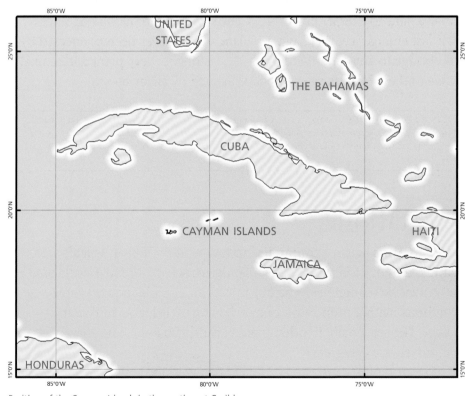

Position of the Cayman Islands in the northwest Caribbean

Geological and biological evidence indicates that the islands last emerged from a period of total submergence two to three million years ago. This was the time that history began for the modern land plants and animals of the Cayman Islands.[1]

Positioned 19 degrees north of the equator, the Islands lie in Holdridge's "subtropical dry forest" life zone. The climate is strongly seasonal, with a sub-humid to humid rainy season between May and November, and a semi-arid to arid dry season between December and April. Average daily minimum-maximum temperatures are 21–28 °C during January and February, rising to 25–32 °C in July and August. Extremes recorded between 1973 and 1987 spanned from 11.2 °C to 36.5 °C. Annual rainfall averages between 1,049 mm and 1,595 mm at various locations on the Islands, with a strong trend to increased rainfall from east to west on Grand Cayman. Occasional hurricanes are a major feature of the climate, exerting a powerful influence on both landforms and biological communities.[2]

Biologically, the Cayman Islands are part of the Greater Antilles, with many of their native plants and animals derived from species on Cuba, Jamaica, and to a lesser extent the Central American mainland. Since their last emergence, the Cayman Islands have never been connected by a land bridge to any other land mass. Ancestors of every plant and animal native to these Islands can only have arrived by air or by sea.

REFERENCES: [1]Jones, B. (2000). Geology of the Cayman Islands. University of Alberta, Edmonton, Canada. [2]Burton, F.J. (1994). Climate and tides of the Cayman Islands. In: *The Cayman Islands: Natural History and Biogeography*, ed. M.A. Brunt & J.E. Davies, pp. 51–60. Kluwer Academic Publishers, Dordrecht, The Netherlands.

Vegetation

Before humans began altering the landscape, the Cayman Islands were blanketed in dry forests, dry shrublands, seasonally flooded wetlands, and extensive mangroves. The dry forests and shrublands are semi-deciduous, with a proportion of trees losing their leaves in the dry season. The seasonally flooded forests and shrublands, and the mangroves, are evergreen.

The dry forests and dry shrublands contain the Cayman Islands' richest terrestrial biodiversity. It is here that the majority of uniquely

Dry forest in the Mastic Reserve, Grand Cayman

Caymanian plants have evolved. Treasuries of the Cayman Islands'
unique natural heritage, these dryland communities are an irreplaceable
capital asset for sustainable nature tourism.

The extensive mangrove communities of Grand Cayman and Little
Cayman provide an impressive array of natural services. They moderate
local rainfall patterns, filter pollutants, contribute nutrients to the marine
ecosystem, serve as a nursery for a wide range of marine life, capture
atmospheric carbon dioxide, adjust to rising sea levels, and protect
coastlines from storms.

These natural assets and services are all now at risk. The loss of the
Cayman Islands' wild forests, shrublands and wetlands can be regarded
as the irreversible liquidation of cultural and economic capital. This is an
accelerating process, sacrificing a unique resource for short-term financial
gains. The loss of these habitats closes options for future human
generations, and now threatens a whole community of Caymanian flora
and fauna with extinction.

CLASSIFICATION OF PLANT COMMUNITIES:

The various natural plant communities that are characteristic of the Cayman Islands were first described and classified by M.A. Brunt in 1984[3]. A more detailed and modern classification is now possible, and is presented separately in full on the CD that accompanies this book. A general overview of the main plant communities is given here.

REFERENCES: [3]Brunt, M.A. (1984). Environment and plant communities. In: *Flora of the Cayman Islands*, G.R. Proctor, pp. 5–58, Kew Bulletin Additional Series X1. The Royal Botanic Gardens, Kew.

FORESTS:

The dry forests of the Cayman Islands range in main canopy height from 4.5–16 m. They have occasional tall emergent trees above the canopy, and an uneven understory layer beneath. They generally occur in areas where the land surface is at least 2 m above the water table. The tree canopy tends to be higher on higher land.

These forests are floristically quite distinct on each of the three islands. The dry forests of Grand Cayman and Little Cayman are classified as lowland semi-deciduous forests, dominated by Red Birch

Red Mangrove forest fringing Jackson's Pond on Little Cayman

Xerophytic shrubland in the east interior of Little Cayman

Dwarfed Red Mangrove forming a shrubland ringed by White Mangroves, in interior Little Cayman

(*Bursera simaruba*) and Cabbage Tree (*Guapira discolor*). On Cayman Brac's elevated plateau, the Bluff, the forests are more drought-adapted and are classified as xeromorphic semi-deciduous forests, dominated by Red Birch and the tree-like cactus *Pilosocereus swartzii*.

Seasonally flooded / saturated semi-deciduous forests are characterised by Royal Palms (*Roystonea regia*), Bull Thatch palms (*Thrinax radiata*), and Mahogany (*Swietenia mahagoni*) on Grand Cayman. On Little Cayman, Poison Tree (*Metopium toxiferum*) is also typical of this forest formation, and Royal Palms are absent. The formation is not present on Cayman Brac.

In wetlands, evergreen semi-permanently flooded forests and tidally flooded mangrove forests occur on peat substrates formed by the plants themselves. These are variously dominated by Black Mangrove (*Avicennia germinans*), White Mangrove (*Laguncularia racemosa*) and Red Mangrove (*Rhizophora mangle*), with Buttonwood (*Conocarpus erectus*) in less saline areas.

Natural coastal herbaceous vegetation and shrubland at the east end of Little Cayman

SHRUBLANDS:

The transition from forest to shrubland occurs where the canopy falls below about 4 m tall. In dry settings, this is accompanied by a distinct change in floristic composition. The dry shrubland communities grow on the eastern ends of all three islands, where rainfall is least. They also occur in other dry land areas where the water table is less than 2 m below the surface, or where the vegetation is stressed by salt spray. Characteristic species include Corato (*Agave caymanensis*) and Wild Cocoplum (*Savia erythroxyloides*).

The semi-permanently flooded forests and the tidally flooded mangrove forests all have shrubland counterparts, which are mainly characterised by reduced canopy height.

Coastal shrublands of Sea Grape (*Coccoloba uvifera*), and a band of specialised coastal shrub species on the seaward side, are a characteristic feature of undisturbed coasts around all three islands. They are only naturally absent in mangrove areas, and on higher elevation coasts with cliffs.

DWARF-SHRUBLANDS:

Mixed evergreen / drought-deciduous dwarf-shrublands are characteristic of the rocky "ironshore" coasts of all three islands. These are low-diversity communities, adapted to the constant salt spray blowing in from the sea.

HERBACEOUS:

Natural herbaceous communities were rare in the Cayman Islands before human disturbance. Areas of sedge and grass wetlands, tidally flooded succulent vegetation, and some sand beach communities are probably the only original herbaceous formations. The grassland communities present in all three islands are a consequence of human activities.

Prostrated Sea Grape forming a dwarf-shrubland on the south-east coast of Cayman Brac

Aerial view of Grand Cayman in 2007, looking over West Bay and the substantially deforested western half of the island, towards the south-east

Deforestation and Endangerment

The human population in the Cayman Islands has been growing exponentially and at times hyper-exponentially since the Islands were first permanently settled in the eighteenth century. Deforestation patterns have been linked to this population growth, meeting escalating demands for housing, transportation, tourism, agriculture and commerce.

The first comprehensive assessment of the Cayman Islands' remaining natural forest cover is based on satellite imagery dating from 1998. At that time, 37% of the land area of Grand Cayman, 19% of Little Cayman, and 26% of Cayman Brac had been converted from natural vegetation into a variety of man-modified landscapes (see maps on pages 24–25). More recent imagery is now being analysed, and it is obvious that deforestation has been ongoing and is accelerating.

Taking past and current trends of human population growth and deforestation and extrapolating them into the future strongly suggests that by the end of this century, all three of the Cayman Islands will be completely deforested. This projection places a high proportion of the Cayman Islands native flora on the endangered species list.

For a long-lived species such as the endemic Cayman Ironwood tree (*Chionanthus caymanensis*), the end of this century is less than a single generation away. New seedlings germinating today are likely to be lost to deforestation before they ever reach maturity. The natural areas which are currently protected, and any which become protected in the near future, will become the last fragments of Caymanian nature, vulnerable oases of biodiversity in a desert of urbanisation.

Former mangrove forest in western Grand Cayman, converted for real estate development

Logwood tree in bloom in eastern Grand Cayman

Alien Invasion

Since the time of original clearance, some areas have been left fallow or under low-intensity agricultural use. New vegetation formations have become established as a result. Many Buttonwood wetlands cleared to make cattle pasture have now stabilised as seasonally flooded grasslands, and low herbaceous perennials are found in saturated pasture. Very few of the grasses are native.

Seasonally flooded areas on Grand Cayman have also been heavily colonised by the invasive, non-native Logwood Tree (*Haematoxylum campechianum*), which forms a near monoculture over extensive land areas. The species was originally introduced to serve as a dye wood.

Soil-blanketed areas, where the original dry forest has been cleared, were originally used for cultivation of crops and fruit trees. In recent years, many of these "grounds" have been turned over to cattle grazing. Short grasslands or medium-tall grasslands (depending on moisture availability) are now established in these zones and in many other heavily disturbed areas. Grasslands may gradually revert to forest, but the process is often set back by dry season fires, and forest species may be overtaken by invasive exotic plant species.

Aggressive spread of the Australian Pine (*Casuarina equisetifolia*) in

African grasses in a traditional pasture, cleared from forest in eastern Grand Cayman

sandy coastal areas has generated a needle-leaved evergreen woodland formation in place of native coastal shrubland. The carpet of fallen *Casuarina* "needles" suppresses germination of other plants. Rampant spread of the Indo-Pacific shrub, Beach Naupaka (*Scaevola sericea*) is also hastening the decline of native beach-ridge vegetation.

The most extreme conversions of native vegetation are in the densely populated areas of western Grand Cayman, where is it not unusual to see manicured, extensively planted landscapes around buildings, with not a single native plant species present.

Protected Areas

Of the 192 threatened plant taxa native to the Cayman Islands, 129 are threatened for one reason only – the destruction of the habitat upon which they depend. Other taxa are also threatened by risks associated with small population size, but these also are primarily threatened by deforestation. Habitat protection is the single most important measure needed to safeguard the majority of the Cayman Islands' native plants from extinction.

Fortunately for conservation planning, threatened plant taxa tend to occur together, in the particularly rich plant communities which have been described and mapped on all three of the Cayman Islands.

On Grand Cayman, the dry forests of North Side and East End, especially the forests of the Mastic area, are particularly rich in endemic and threatened plants. Fragments of forest remaining in George Town and Spotts retain other species which were once widespread in the wetter, western districts. The dry shrublands in the far east of Grand Cayman form a second major centre of diversity for uniquely Caymanian plants.

On Little Cayman, the Central Forest which lies north of the Tarpon Lake wetland and the dry shrublands on Sparrowhawk Hill contain the majority of Little Cayman's known plant biodiversity. Sandy coastlines, especially at Point of Sand and facing Bloody Bay, retain near-natural coastal shrubland and forest of a kind which has been lost from much of the Caribbean.

On Cayman Brac, the dry forests of the central Bluff and the dry shrublands at the elevated eastern end of the island are the two richest areas, and they support the majority of Cayman Brac's unique and characteristic plants.

These same areas are also of outstanding importance to the Cayman Islands' unique fauna. It is no coincidence that the efforts of the National Trust for the Cayman Islands, and of the Cayman Islands Department of Environment, are strongly focused on protection of these same areas.

Major terrestrial protected areas in the Cayman Islands currently include the Mastic Reserve and Salina Reserve on Grand Cayman, the Booby Pond Nature Reserve on Little Cayman, and the Brac Parrot

Australian Pines and Beach Naupaka invading Grand Cayman's south coast after hurricane Ivan

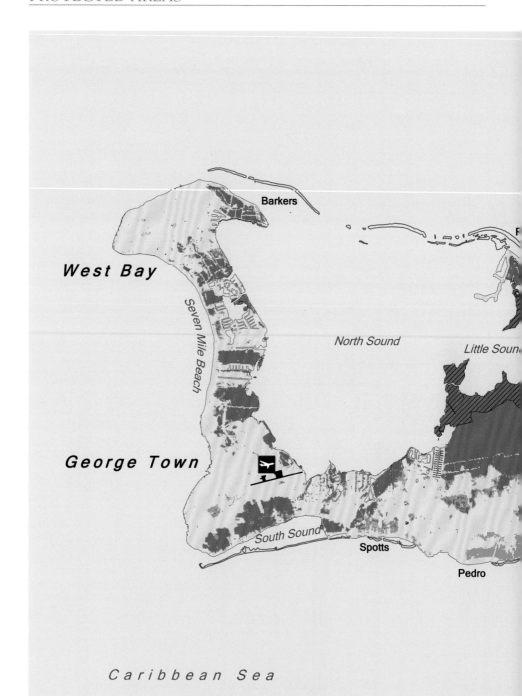

Grand Cayman: vegetation cover and terrestrial protected areas

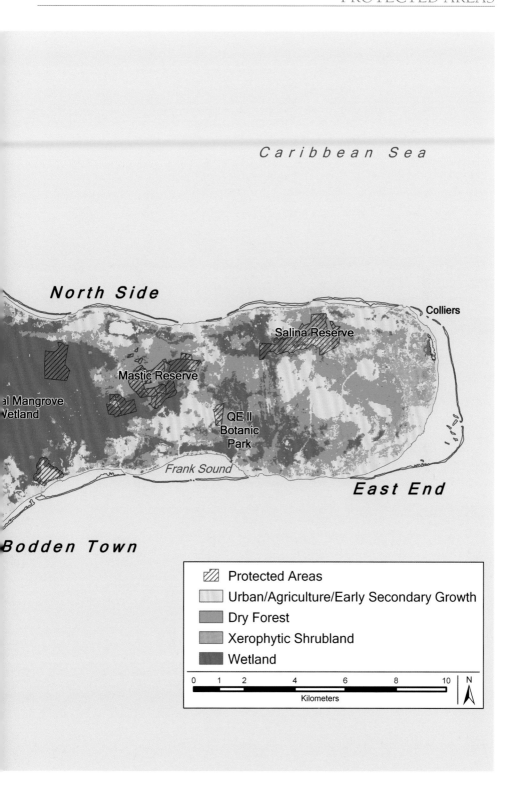

Caribbean Sea

North Side

Colliers

Salina Reserve

Mastic Reserve

al Mangrove
Wetland

QE II
Botanic
Park

Frank Sound

East End

Bodden Town

Protected Areas
Urban/Agriculture/Early Secondary Growth
Dry Forest
Xerophytic Shrubland
Wetland

0   1   2   4   6   8   10   N

Kilometers

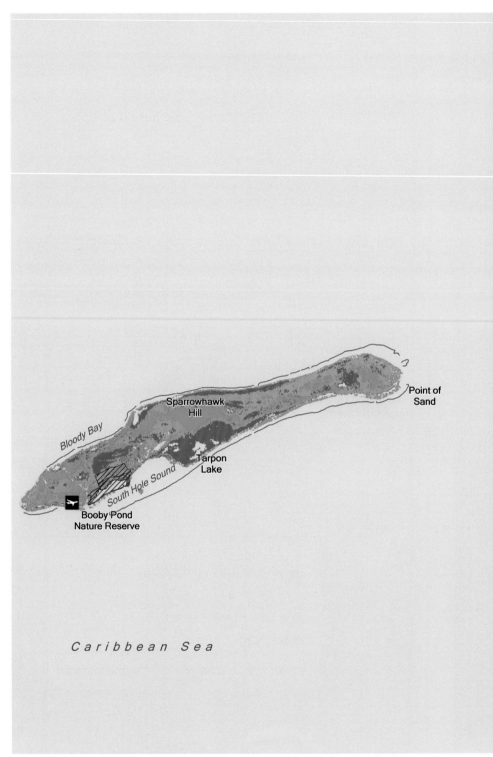

Little Cayman (left) and Cayman Brac (right): vegetation cover and terrestrial protected areas

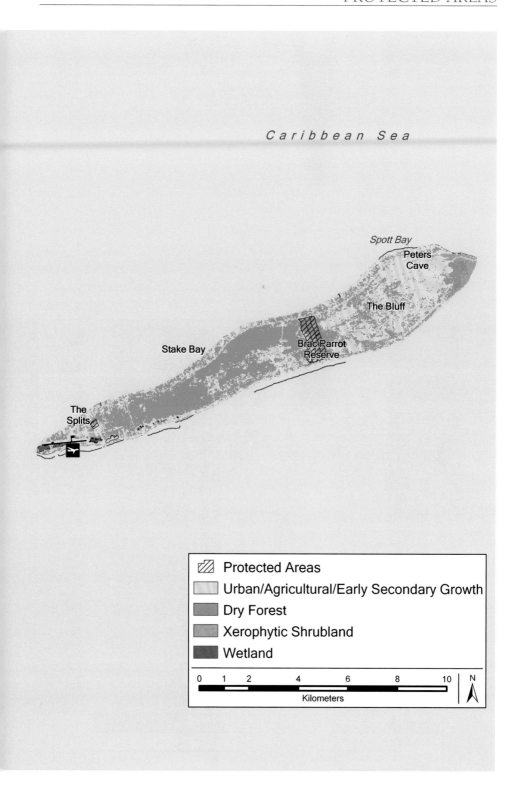

Caribbean Sea

Spott Bay

Peters Cave

The Bluff

Stake Bay

Brac Parrot Reserve

The Splits

Protected Areas

Urban/Agricultural/Early Secondary Growth

Dry Forest

Xerophytic Shrubland

Wetland

| 0 | 1 | 2 | | 4 | | 6 | | 8 | | 10 | N |

Kilometers

Cladium sedge marsh in the Salina Reserve, Grand Cayman

Reserve on Cayman Brac (see maps on pages 24–27). All are protected under the National Trust for the Cayman Islands Law.[4] Fragments of Grand Cayman's Central Mangrove Wetland have also come under protection, by the Trust and through the Environmental Zone of the Cayman Islands Marine Park system. As of 2007, these and other smaller protected areas comprised 7.1% of the Cayman Islands' land area.

This represents very significant progress. However, only half the target area for the Mastic Reserve is protected, Grand Cayman's dry shrubland ecosystem is almost all completely unprotected, and critically important fragments of forest in George Town and Spotts are likely to be lost at any time.

The Brac Parrot Reserve now protects 112 ha of the Brac's finest forest, but this is still an insufficient area to support many wide-ranging species. The unique dry shrublands at Cayman Brac's east point, and the coasts and pristine Central Forest of Little Cayman, remain entirely unprotected.

Grand Cayman	Little Cayman	Cayman Brac	Total
8.0%	5.6%	3.2%	7.1%

Table 2: Percentage of land area under environmental protection, as of 2007

Clearly, if the unique plants, animals and environments that define the very nature of the Cayman Islands are to survive, steps to protect these vulnerable areas before they are lost to deforestation are essential, and of the utmost urgency.

REFERENCES: [4]National Trust for the Cayman Islands Law 22 of 1987 (1997 Revision).

Notes on methods

Threatened Plants of the Cayman Islands is a Red List assessment of the Cayman Islands' native plants. It is the first formal assessment of the conservation status of all the plants believed to be part of the original flora of these islands. Each species is classified according to the current Red List system, using globally standardised categories developed by IUCN – the World Conservation Union – to evaluate the risk of extinction of a species.[5]

Data for these assessments came largely from a database of plant species abundances by site, which was generated by an extensive forest biodiversity survey carried out by the author in 2000. These data were combined with data from the Cayman Islands' national herbarium (CAYM), which is managed by the National Trust for the Cayman Islands. The author's own field observations, and field observations contributed by several other local naturalists, supplemented these sources.

A simple model relating deforestation to documented human population growth on each island, with adjustments made for relevant historical events such as a period of commercial hardwood extraction, was constructed. Curve-fitting was then used to create a projection to the year 2100, and this was critically examined through alternative scenario discussions before being accepted as the appropriate projection for Red List assessment purposes. Existing and credibly proposed protected areas were assumed to remain protected from deforestation throughout the future.

The model was used to estimate past population changes, and to forecast future population trends for each native plant species, starting from an estimated population size in 2000 that was derived from the biodiversity survey data set. The Red List assessment was then carried out using RAMAS® Red List software.

Where quantitative population estimates were not available, Red List assessments have been made on the basis of other criteria wherever possible, using whatever objective data could be brought to bear. Plants species and varieties have been classed data-deficient where there is insufficient evidence to assign them to any other Red List category.

Taxonomic nomenclature in this book follows that of G.R. Proctor's *Flora of the Cayman Islands*, second edition[6], wherever possible, but recognising that several of the taxa presented in this book are currently under taxonomic review.

REFERENCES: [5]www.redlist.org & www.iucn.org [6]Proctor, G.R. (in press). *Flora of the Cayman Islands*, second edition. The Royal Botanic Gardens, Kew.

Key to the plant description pages

The plant description pages are divided into two sections: endemics and near-endemics. The format for the pages is the same for both sections. Within each section, plants are organised alphabetically by family name.

Plant family – The relevant plant family name runs across the top of each page.
Scientific name and **Authority**
var. and **Authority** (if applicable)
Common name – If applicable, the most common name used in the Cayman Islands is listed.
Status – **Red List Status**.
Red List Criteria – The standard IUCN abbreviation which indicates the reasons why a listed plant is threatened.
Distribution – A description of the geographic range where the plant occurs naturally.
Threats – A summary of the threats to the survival of the plant.
Needs – Top-priority conservation action(s) needed.
Ecology – Ecological and other information of interest.

The **Plant family** is a scientific Latin name for groups of related species. The formal **Scientific name** consists of a genus name, followed by a species name (both in Latin), and then the name of the **Authority** who gave those names (often abbreviated). Some plants also occur in distinct varieties, in which case a third Latin name follows after the abbreviation "**var.**", with another **Authority**. The genus, species, and variety names provide scientists with an unambiguous, standardised way to label each recognisably different kind of plant.

For plants which have common names in use within the Cayman Islands, the most widely used name is given. Unlike scientific names, **Common Names** are often applied to several different kinds of plant, and any one kind of plant may have several common names.

The **Red List Status** categorises the degree to which each plant species or variety is at risk of extinction within the Cayman Islands.

Extinct (EX) means there is no reasonable doubt that the last individual died.

Extinct in the Wild (EW) means that the taxon is known only to survive in cultivation or captivity.

Critically Endangered (CR) means facing an extremely high risk of extinction in the wild.

Endangered (EN) means facing a very high risk of extinction in the wild.

Vulnerable (VU) means facing a high risk of extinction in the wild.

Near Threatened (NT) means likely to become Vulnerable or worse, in the near future.

Least Concern (LC) applies to plants which are neither threatened nor near-threatened.

Data Deficient (DD) means that there is insufficient information available to assess whether a plant is threatened or not.

The **Red List Criteria** code that appears beneath the **Red List Status** indicates why a particular taxon is threatened. Code elements prefixed by the letter "A" relate to past and future reductions in population size. Code elements beginning with "B" relate to limited geographic range or limited area of occupancy. "C" and "D" codes address threats associated with small population size, while "E" codes (not used here) are applied to mathematical modeling of the probability of extinction. A full definition of all these criteria and their codes is published at www.redlist.org.

In the **Needs** paragraph, "propagation" refers to reproduction of plants for conservation management, while "cultivation" refers to the production of plants for commercial landscaping and private gardens.

PLANT DESCRIPTIONS
1. The Endemic Flora

The strictly native flora of the Cayman Islands consists of 415 taxa (species and varieties). Of these, twenty-one species of higher plants are now known to be unique (endemic) to the Cayman Islands. A further eight species are present as endemic Caymanian varieties, and one endemic Caymanian species has two endemic varieties. Overall, there are 29 uniquely Caymanian plants.

Ten of these species and varieties have only been discovered, or recognised to be distinct, over the course of the past decade. Almost certainly, the list of the endemic flora presented here is still incomplete.

Continuing research is needed to fully document the unique botanic heritage of the Cayman Islands. It is also likely that the large-scale deforestation of much of the Cayman Islands may have caused some unique plant species to be lost before we ever knew they existed.

Agave caymanensis Proctor

Common name – CORATO

Status – Vulnerable
VU A2c+3c+4

Distribution – Endemic to the Cayman Islands. Found only on Grand Cayman, Little Cayman and Cayman Brac, in dry shrubland settings at all elevations, especially at the drier eastern ends of each island where it is often a dominant member of the plant community.

Threats – Habitat conversion for human uses, occurring over time frames notably shorter than the generation time for this very slow-growing species.

Needs – Habitat protection.

Agave caymanensis flowers

Agave caymanensis habit

Ecology – This impressive species gradually forms a short trunk, which is deeply clothed in dead leaves. Corato can grow to as much as 4 m tall and 3 m wide with a rosette of massive, succulent, thorn-rimmed leaves. Like all Agaves, it is "monocarpic", meaning it flowers only once in its lifetime. It is sometimes known as the "Century Plant", hinting at great age. What eventually triggers each plant to flower remains unknown, but those that do flower in a given year are usually quite synchronous.

The flower spike starts to grow from the centre of the rosette in January. The spike may reach 6 m or more in height before producing an extravagant mass of orange-yellow, nectar-rich flowers, usually in March. Bananaquits and bees feast on the nectar and pollinate the flowers by day, and it is likely that bats and hawk moths continue to do so by night.

As the seed capsules mature and burst, the inflorescence also develops numerous "bulbils" – miniature plants cloned off their parent, which fall ready to grow. After this orgy of reproduction, the parent plant dies and gradually dries into a carcass which takes many years to decompose. The hollow core which remains after the dead flower spike falls is sometimes used as a refuge by young rock iguanas.

Pilostyles globosa (S. Wats.)
Hemsley
var. *caymanensis* Proctor

Status – Least Concern

Distribution – A variety unique to the Cayman Islands. Widespread, mainly in undisturbed dry shrubland and semi-deciduous forest. The species is also known from Central America and Jamaica.

Threats – Habitat conversion is reducing the population of this parasitic plant's host tree, but the trend is offset by some re-colonisation of disturbed areas.

Needs – Research (reproduction), and habitat protection.

Ecology – This remarkable plant is an endoparasite which grows largely inside the stems of the Bull Hoof tree (*Bauhinia divaricata*), taking all its nutrients from its host. The buds and flowers are the only part visible outside. An infected Bull Hoof stem is usually encrusted with dozens of buds, flowers and scars from old *Pilostyles* flowers, at various stages of development. The host tree does not seem to be severely affected by the parasite: it grows, flowers and sets seed despite having the parasite in almost every

Pilostyles globosa var. *caymanensis* on stem of *Bauhinia divaricata*

young stem. The bark of the Bull Hoof grows over the *Pilostyles* flower scars as older stems thicken into branches, and the infection seems to migrate into younger stems as they develop. How those flowers are pollinated, and how the seeds are dispersed, has never been studied.

Pilostyles globosa var. *caymanensis* flowers

Pectis caymanensis var. *caymanensis* flowers

Pectis caymanensis (Urb.) Rydb.
var. robusta Proctor
var. caymanensis Proctor

Common name – TEA BANKER

Status – Critically Endangered
CR A2acd; B1ab(i,ii,iii,iv,v)+2b(i,ii,iii,iv, v), both varieties

Distribution – Occurs in two varieties. The variety *Pectis caymanensis* var. *robusta* Proctor is unique to Grand Cayman. The variety *P. caymanensis* var. *caymanensis* occurs on the three Cayman Islands and Cuba.

Threats – The few recently confirmed locations where this species grows are all coastal. Here, populations have been eradicated by real estate development, and recently by severe coastal erosion during hurricanes Gilbert, Ivan and Wilma in 1988, 2004 and 2005, respectively. Historically, harvesting to make a tea probably also contributed to the decline. Information on the past and present occurrence of this species in the Cayman Islands is very sparse, and the status assessments for both varieties would benefit from targeted surveys. The variety *robusta* has not been recorded in recent years, and may even be extinct.

Needs – Research (status, distribution), *ex situ* propagation, seed banking, awareness, legal protection, and habitat protection.

Ecology – Tea Banker is a small, inconspicuous, mat-like herb, spreading out from a tap-root. It has a strong aroma, often released when the plant is accidentally crushed underfoot. On close examination the leaves are distinctive: they have a hair at the tip and hairs along the leaf margins towards the base, and the underside has rows of conspicuous glands. The stems are often suffused with pink, and the small flowers have yellow ligules. As its name suggests, variety *robusta* is a larger, coarser form.

Pectis caymanensis var. *caymanensis* habit

Verbesina caymanensis
Proctor

**Status – Critically Endangered
CR B1ab(v)+2ab(v)**

Verbesina caymanensis flowers

Distribution – Unique to Cayman Brac, and only found on the cliffs of the Bluff over Spot Bay, where the near-vertical rock face is oriented due north. The species can be seen from both Peter's Road and Big Channel Bluff Road, where these traditional trails ascend the cliff. The rock face is in permanent shade except for the few summer months when the sun's azimuth passes north of the island.

Threats – The entire world population of this species occurs only in this single location, and so could be rendered extinct by any locally catastrophic event. The non-native shrub *Tecoma stans* and herb *Bryophyllum pinnatum* are displacing *Verbesina caymanensis* on the Peter's Road ascent, a situation compounded by occasional trail clearance and burning.

Needs – *Ex situ* propagation, seed banking, awareness, legal protection, and habitat protection.

Ecology – This is a small shrub which roots into rock crevices. The leaves are somewhat rough-textured and have serrations on the leaf margins, especially towards the tip. The flowers are white. Little is known of the biology of this species. Despite being restricted to a very specific habitat in the wild, it was grown successfully in cultivation in the Queen Elizabeth II Botanic Park, on Grand Cayman, for a number of years.

Verbesina caymanensis habit

Cordia sebestena L. var. *caymanensis* (Urb.) Proctor

Common name – Broadleaf

Status – Vulnerable
VU A3bc+4bc

Distribution – A variety unique to the Cayman Islands. The species has a broad neotropical distribution.

Threats – The most abundant stands of this variety in the Cayman Islands are in coastal beach ridges, where its population has been severely impacted by accelerating real estate development. Invasive alien species such as Australian Pine (*Casuarina equisetifolia*) and Beach Naupaka (*Scaevola sericea*) are also displacing Broadleaf from coastal settings. Less dense populations also occurs in drought-deciduous forests and xerophytic shrublands on all three islands. Ongoing habitat conversion is also taking a toll in those areas. Despite the ease with which this

Cordia sebestena var. *caymanensis* fruit

variety can be cultivated, the economics of the landscaping industry are such that the widespread variety from Florida is imported and planted instead of the native, and equally attractive, variety *caymanensis*. Hybridisation between the two seems likely to follow.

Needs – Seed banking, cultivation, awareness, legal and policy measures against imports, and habitat protection.

Ecology – The variety *caymanensis* is distinguished by its deep red flowers, and large, serrate leaves. The leaves were used traditionally to polish turtle shells and drinking glasses. The fibrous white fruits are consumed by native rock iguanas. Broadleaf can be grown easily from seed, and grows rapidly in good soil. It typically forms a slender, open-crowned small tree which flowers frequently in response to rain.

Cordia sebestena var. *caymanensis* flowers

37

Hohenbergia caymanensis

Britton ex L.B. Smith

Common name – OLD GEORGE

Status – Critically Endangered
CR A3c+4c; C1+2a(ii)

Distribution – Unique to western Grand Cayman. Restricted to rather humid settings. This species is only known from undisturbed forest and shrubland fragments in southern George Town. A second population has been established as a conservation collection in the Queen Elizabeth II Botanic Park.

Threats – George Town's remaining fragments of natural forest are disappearing very rapidly, and *Hohenbergia caymanensis* is disappearing with them. The land on which this species grows is now highly valued for real estate. Most of the areas where *Hohenbergia* used to be abundant are now covered with housing and landscapes of non-native plants.

The conservation collection in the Queen Elizabeth II Botanic Park is derived from few locations and does not represent the genetic diversity still present in the wild. The species has flowered occasionally in the Botanic Park but has not produced a new generation, so there is no guarantee that the translocation will be successful in the long term. Preservation of this species in its natural habitat should be the highest priority.

Hohenbergia caymanensis habit

Needs – Research (reproduction, distribution), *ex situ* propagation, seed banking, cultivation, awareness, legal protection, and habitat protection.

Ecology – In the wild, *Hohenbergia caymanensis* is often an epiphyte, growing on tree branches, as well as epilithic, growing on the "cliff rock" surface. Where it survives, it can form quite dense populations. It is found in dense, humid forest close to wetlands. This is a large species with leaves to 1 m long. The branching inflorescence arches from between the leaves, and develops fruit capsules about 4 cm long with multiple seeds. Mature individuals also reproduce clonally, by budding off new plants from the base. The leaves are minutely toothed along the margins. Rainwater and leaf litter collects in the leaf bases, forming a miniature aquatic ecosystem and a humid refuge for native frogs.

Hohenbergia caymanensis flower spike

Consolea millspaughii

(Britton) A. Berger

var. *caymanensis* Areces

**Status – Critically Endangered
CR C2a(ii)**

Distribution – A variety unique to Cayman Brac, and formerly Little Cayman. The extent to which this variety is distinct from *Consolea moniliformis* on Cuba is unclear. On Cayman Brac it occurs in the easternmost shrublands on top of the Bluff. It is also scattered sparsely along the south bluff edge, at both the top and the bottom of the cliffs. *Consolea millspaughii* var. *caymanensis* used to occur in eastern Little Cayman, but has apparently died out there.

Threats – The original bluff-brink population on Cayman Brac was probably severely reduced by clearing and burning, historically done to aid harvesting of Brown Booby eggs (these are now protected). The reasons for this species' disappearance from

Consolea millspaughii var. *caymanensis* flower

Little Cayman are entirely unclear. Preliminary studies indicate that the individuals on Cayman Brac, as well as one specimen in cultivation from Little Cayman, may all be males! [7] No fruiting specimens have ever been seen, and many individuals may be clones of single plants. Some specimens are injured severely by mining caterpillars, probably of the notorious *Cactoblastis* moth.

Needs – Research (taxonomy, reproduction), *ex situ* propagation, seed banking (if possible), awareness, legal protection, and habitat protection.

Ecology – This is a visually attractive cactus, which produces abundant bright red flowers. However, its spines are extraordinarily sharp and will pierce and adhere to skin and clothing, to such an extent that few people will tolerate it in their gardens. It is this ability to stab and stick that probably accounts for the dispersal of small pads, which hitch rides on birds and people. *Consolea millspaughii* var. *caymanensis* is adapted to very dry, exposed situations. Old specimens shed numerous small pads, which root readily and are very easy to transplant.

Consolea millspaughii var. *caymanensis*

REFERENCES: [7]Strittmatter, L. (2005). Pers. comm.

Epiphyllum phyllanthus var. *plattsii* habit

Epiphyllum phyllanthus

(L.) Haw.

var. *plattsii* Proctor

Status – Critically Endangered
CR B1ab(i,ii,iii,v)c(i)+2ab(i,ii,iii,v)c(i); D

Distribution – A variety apparently unique to Cayman Brac. It is known only from a single colony, towards the west end of Cayman Brac's bluff. A report of a second colony from the Green Lane area, to the far east of the bluff, has not been confirmed despite considerable searching. The species is distributed from Central to South America.

Threats – As far as we know, the entire world population of this unique Cayman Brac variety grows in just one location, in a colony about 70 m in diameter, and with a road constructed right through its centre. Cayman Brac's bluff-top forests are being cleared piecemeal for farming, and most recently, for housing development. Should either land use strike that specific spot, *Epiphyllum phyllanthus* var. *plattsii* could be extinct in the wild in a matter of minutes.

Needs – Research (taxonomy, status, distribution), *ex situ* propagation, seed banking, awareness, legal protection, and habitat protection.

Ecology – Cayman Brac's *Epiphyllum* scrambles over rocky terrain in the shade of a low dry forest canopy, rooting into leaf litter in rock crevices. This unusual, spineless cactus has flattened green stems with wavy margins. The nocturnal white flowers bud out from the edges of the stems, later forming bright purple spindle-shaped fruits.

The presence of this cactus at a single spot on Cayman Brac is a mystery, especially since this genus is not known to grow wild anywhere else in the West Indies. Our understanding of its distribution and status may change as more information comes to light.

Epiphyllum phyllanthus var. *plattsii* fruit

Crossopetalum caymanense
Proctor

Status – Least Concern

Crossopetalum caymanense fruits

Distribution – A species unique to the Cayman Islands, occurring on all three islands. This is a small dry forest shrub, growing mostly in the shade on rocky substrates and in soil patches, never extremely common, but present almost wherever there is native tree cover.

Threats – Ultimately, the survival of this species in the face of the progressive deforestation of the Cayman Islands will depend on establishing and maintaining protected forest areas. This is a fast-maturing species, however, and over its next few generations there are no imminent threats of catastrophic decline.

Needs – Habitat protection.

Ecology – Delicate arching stems, tiny deep purple flowers, and conspicuous bright red berries make this an attractive denizen of the forest floor. A form at the eastern end of Cayman Brac has much larger leaves than specimens on the rest of the Brac and the other two islands. *Crossopetalum caymanense* grows easily from seed, and flowers and fruits frequently throughout the year. It has potential as a garden plant.

Crossopetalum caymanense flowers

Terminalia eriostachya var. *margaretiae* flowers

Terminalia eriostachya

A. Rich

var. *margaretiae* Proctor

Common name – BLACK MASTIC

**Status – Critically Endangered
CR A3bc+4bc**

Distribution – A variety unique to Grand Cayman. It is now restricted to the dry forests of North Side and East End. The species also occurs in Cuba.

Threats – Historically, this tree was apparently widespread in Grand Cayman, but by 1800, it was being harvested to the point of extinction for its dark, ebony-like heartwood. The remnant population survives mainly in the Mastic forest, and is now partially protected in the Mastic Reserve. This is a long-lived species which occurs naturally at low density. It therefore requires large forest areas to maintain a viable population.

Needs – Propagation, seed banking, cultivation, awareness, legal protection, and habitat protection.

Ecology – Black Mastic is a large tree, forming a broad crown in the upper forest canopy. It does not bloom every year, but when it does, it becomes festooned with dangling inflorescences of yellow flowers, which Cayman parrots love to eat. The seeds are embedded in flat, papery disks, and are dispersed by wind. They germinate readily in light soil, but the roots are intolerant of waterlogging. In rocky forests, this is a slow-growing species, but if planted in deep soil, it grows relatively rapidly and can make a fine landscape specimen tree.

Terminalia eriostachya var. *margaretiae* seeds

Argythamnia proctorii

Argythamnia proctorii
Ingram

Common name – CAYMAN SILVERBUSH

Status – Least Concern

Distribution – A species unique to the Cayman Islands. Widespread in forests of all three islands, this small perennial shrub is especially abundant on Cayman Brac, where it grows in dappled shade under trees.

Threats – Like all the Cayman Islands' native flora, *Argythamnia proctorii* populations are declining rapidly as natural forest cover is lost. However, this is a short-lived, prolific-seeding species which is capable of re-colonising second growth woodlands.

Needs – Habitat protection.

Ecology – Cayman's *Argythamnia* is an erect, wiry-stemmed species which grows to about 1.5 m tall. The youngest leaves are silvery beneath. The species is monoecious (bearing separate male and female flowers on the same plant). Seed capsules are produced year-round. Once ripe, the capsules dry and explode, scattering the seeds.

Argythamnia proctorii flowers

Chamaesyce bruntii Proctor

**Status – Critically Endangered
CR B1b(ii,iii)**

Distribution – A species only recorded from Little Cayman. It grows in open areas on sandy beach ridges, and other disturbed habitats near the coast.

Threats – Although locally common, this species is restricted to the 28.5 km² island of Little Cayman. There, it occurs in coastal habitats which are beginning to come under pressure from real estate development. No estimates of its population size have been made to date.

Needs – Research (status and distribution), *ex situ* propagation, seed banking, and habitat protection.

Chamaesyce bruntii flowers

Ecology – This is a perennial herb with a taproot, usually growing flat on the sand. It is similar to the much more widespread *Chamaesyce prostrata*, which unlike this species has minutely hairy stems. Like all its kin, *Chamaesyce bruntii* oozes white sap from broken stems and leaves.

Chamaesyce bruntii

Phyllanthus caymanensis

Phyllanthus caymanensis

Webster & Proctor
Status – Vulnerable
VU D1+2

Distribution – A species unique to the Cayman Islands. Originally believed to be endemic to Cayman Brac and Little Cayman, it has also been discovered at two locations in the dry shrublands of Grand Cayman's

Phyllanthus caymanensis flowers

east interior. There, it is highly restricted to relatively undisturbed soil zones.

Threats – The global population of this species is now very small, perhaps of the order of 1,000 individuals, distributed among a restricted number of sites on the three Cayman Islands. Forest clearance could lead to the species becoming critically endangered in a very short period of time.

Needs – Habitat protection.

Ecology – *Phyllanthus caymanensis* is a low shrub, with an elegant, open branching habit. It grows up to 2.5 m tall, but often much less. At first glance it appears to have compound leaves, but true to the Euphorbiaceae family's tendency to unusual growth forms, these are actually deciduous branchlets, with simple leaves. It is a facultatively drought-deciduous species, shedding leaves and branchlets in the dry season to a varying degree, depending on rainfall patterns each year.

Salvia caymanensis
Millsp. & Uline

Status – Critically Endangered
CR B1ab(i,ii,iii,iv,v)c(ii,iv)+2ab(i,ii,iii,iv,v)
c(ii,iv)

Distribution – Only ever recorded in
Grand Cayman, with historical collections
from West Bay, the Seven Mile Beach, and
the Frank Sound area. Now known only
from road verges along the Queen's
Highway.

Threats – This species was lost and thought
to be extinct between 1967 and 2006, but
was recently rediscovered in two widely
spaced locations on Grand Cayman. One

Salvia caymanensis

group, on the Queen's Highway, is a perfect
match to the herbarium specimens collected
in the 1960s. The other group, on South
Church Street, is sufficiently different that it
may represent another species altogether.
Since its rediscovery, however, it has been
lost from the wild again. Fortunately, a few
specimens are now in cultivation.

Needs – Research (taxonomy), seed
banking, propagation, reintroduction to the
wild, and habitat protection.

Ecology – The only recent records for this
(or these) species are on periodically mowed
roadside verges, and in an old sand garden
(now lost). Historically, *Salvia caymanensis*
was apparently a species of sandy coasts,
thriving in clearings created by storms. Its
population has probably always fluctuated
with the availability of suitable habitat, and
it is possible the seeds can remain dormant
in the ground for long periods.

Salvia caymanensis flowers

Dendropemon caymanensis
Proctor

**Status – Critically Endangered
CR C2a(i,ii)**

Distribution – A species known only from Little Cayman. The type specimen was collected by G.R. Proctor in or near the Central Forest, and a second collection by Proctor was made in the same area in 1991.

Threats – Since 1991, surveys of the Central Forest on Little Cayman have not yielded any new records of this parasitic plant. Indeed its recorded host plants appear to be very scarce in this area. It seems likely that *Dendropemon caymanensis* exists as a very small population indeed, at a single, highly restricted and unprotected locality. Bulldozer clearance lines through parts of the Central Forest are testament to how a single event could cause the extinction of this species at any time.

Needs – Research (reproduction, ecology, status, distribution), *ex situ* propagation, seed banking, awareness, legal protection, and habitat protection.

Dendropemon caymanensis, an artists impression

Dendropemon caymanensis, herbarium specimen

Ecology – *Dendropemon caymanensis* is known to be a parasite of the Headache Bush (*Capparis cynophallophora*) and Black Candlewood (*Erithalis fruticosa*), and has been recorded on other un-named species. No photographs of the living plant or cultivated specimens, exist. No precise locations of wild plants are mapped. The only visual representations possible here are a photograph of a dried herbarium specimen, and an artists impression derived from it.

Surveys to relocate this species, map its distribution, identify all its host plants, and assess its population size are a matter of some urgency.

Pisonia margaretae
Proctor

Status – Critically Endangered
CR C1+2a(i,ii); D

Distribution – A species known only from the Spotts area of Grand Cayman. A duplicated population is in cultivation in the Queen Elizabeth II Botanic Park, also on Grand Cayman.

Pisonia margaretae flowers

Pisonia margaretae

Threats – This species apparently only ever existed as a very small population on the high land at Spotts. Its numbers were probably halved by the construction and subsequent widening of the main coastal highway. Road verge clearance continues to impact the few remaining specimens, and future road widening and housing developments seem certain to cause near or total extinction of this species in its original location. Unless any additional subpopulations are discovered, the perilous future of *Pisonia margaretae* may soon hinge on management of the conservation collection in the Queen Elizabeth II Botanic Park, which includes root cuttings taken from every one of the 40 individuals known in the wild.

Needs – Research (reproduction), *ex situ* propagation, seed banking, awareness, legal protection, and habitat protection.

Ecology – This is a profusely branching, woody shrub with large leaves. It tends to spread clonally by root suckers, and grows well in cultivation. Specimens in the Botanic Park have set viable seed annually, but to date, no wild seedlings have resulted. The small, spindle-shaped seeds are covered in hooks and adhere to clothing – and presumably, to feathers and fur. With such potentially mobile seeds, the highly restricted distribution of this species is rather mysterious.

Chionanthus caymanensis
Stearn

Common name – IRONWOOD

Status – Endangered
EN A3bc+4bc

Distribution – A species unique to the Cayman Islands. On Cayman Brac and Grand Cayman, it is relatively common in elevated dry forests, and in some areas is one of the dominant tree species. It is less abundant on Little Cayman.

Threats – Deforestation for agriculture, housing, land speculation and other human uses are occurring over time frames notably shorter than the generation time of this very slow-growing species. The largest mature Ironwood trees in Cayman's dry forests are likely to be well over a century old. Historically, Ironwood was also cut selectively for use in traditional house construction.

Needs – Habitat protection.

Ecology – This is one of the Cayman Islands' forest canopy trees, developing a gnarled, fluted and hollow trunk as it ages. The wood is brittle and exceptionally hard and dense, blunting woodworking tools and sinking in water.

Chionanthus caymanensis

Chionanthus caymanensis fruit

Ironwood may pass years with little or no reproduction before unpredictably setting a heavy crop of small purple berries. These are feasted upon by birds and rock iguanas. All the Ironwood trees in the same district tend to fruit synchronously. The seeds germinate readily and the tree can be cultivated easily, provided it is planted well above the water table and the grower has a good deal of patience!

Dendrophylax fawcettii
Rolfe

Common name – Ghost Orchid

Status – Critically Endangered
CR A2abcd+4abcd

Distribution – A species known from Grand Cayman only. Small but intact populations are present in several areas of East End and North Side. Some individuals also survive in George Town's scattered forest remnants. This orchid occurs in humid forests, downwind of seasonally flooded wetlands.

Threats – The higher rainfall of Grand Cayman's former western forests must once have provided extensive habitat for this orchid. Only tiny fragments of forest remain in western Grand Cayman, so the future of the Ghost Orchid now depends on its scattered eastern populations. Accelerating deforestation is projected in those areas too. Minimal numbers of Ghost Orchids are present in protected areas.

Dendrophylax fawcettii

Dendrophylax fawcettii

Needs – Research (pollination), *ex situ* propagation, seed banking, awareness, legal protection, and habitat protection.

Ecology – Ghost Orchids are all roots: they have no leaves, and practically no stem. The roots radiate out from a central point, like the legs of a spider. Adjacent plants tangle together in a web, clinging to tree branches and surface rocks. The flower is pure white, with a long nectar tube, born on a long slender stem. The flower structure suggests they may be pollinated by hawk moths.

Some of George Town's vanishing Ghost Orchids have been rescued from land clearance and translocated to the Queen Elizabeth II Botanic Park, where a proportion appears to be growing well. The translocated plants set viable seed when hand pollinated but do not appear to be successfully reproducing unaided. Such translocations will not be self-sustaining if the relevant pollinator species are missing.

Encyclia kingsii
(C.D. Adams) Nir

Status – Critically Endangered
CR A2abcd+3bc+4bc; C1+2a(i,ii); D

Encyclia kingsii

Encyclia kingsii flowers

Distribution – A species unique to Little Cayman and Cayman Brac. Distributed widely, but extremely scarcely, in dry forests.

Threats – This species is so rare, and distributed so scarcely through the forest, that extensive areas would have to be protected in order to conserve a viable population in the wild. Ongoing deforestation, together with pressure from unscrupulous collectors and the intrinsic vulnerability of very small, localised populations, combine to place *Encyclia kingsii* at an extremely high risk of extinction.

Needs – Research (pollination), *ex situ* propagation, seed banking, awareness, legal protection, and habitat protection.

Ecology – Epiphytic on trees or sometimes on fallen logs, *Encyclia kingsii* produces a fairly conspicuous panicle of small, predominantly yellow flowers. The stems of the panicle have extra-floral nectaries, possibly serving to attract ants, but the pollinator of this unique, delicate orchid is not known.

Myrmecophila thomsoniana var. *thomsoniana*

Myrmecophila thomsoniana
(Rchb. f.) Rolfe

var. *thomsoniana*
var. *minor* (Strachan ex Fawc.)
Dressler

Common name – Banana Orchid

Status – Endangered
EN A3bc+4bc – both varieties, and the species as a whole

Distribution – A species unique to the Cayman Islands. The variety *thomsoniana* occurs on Grand Cayman, and var. *minor* on Little Cayman and Cayman Brac. Banana Orchids are common throughout dry forest and shrubland habitats on all three islands.

Threats – In the past, commercial collection for international trade pressured the Banana Orchid population, especially on Cayman Brac. Such trade is now illegal and has ceased. Accelerating habitat destruction is now the major threat.

Needs – Habitat protection.

Ecology – The Banana Orchid is the Cayman Islands' national flower. The Sister Isles' variety differs most obviously from the Grand Cayman form in having yellow, rather than cream, flowers. Flowers normally appear in April to June. Individual Banana Orchid plants appear to be quite long-lived. As the pseudobulb grows ahead, the trailing structure dies behind, and hence the orchid slowly migrates along the tree trunk or branch.

Banana Orchids have inherited hollow pseudobulbs with entrances, and extra-floral nectaries on the flower stalks, from ancestors on the Central American mainland. There, the pseudobulbs are occupied by stinging ants which protect the orchid, and in turn gain food and shelter. In the Cayman Islands, the pseudobulbs are only occasionally occupied by ants, and small *Anolis* lizards are the most frequent visitors to the nectaries!

Myrmecophila thomsoniana var. *thomsoniana*

Myrmecophila thomsoniana var. *minor*

Coccothrinax proctorii

Coccothrinax proctorii
R.W. Read

Common name – SILVER THATCH

Status – Endangered
EN A3bc+4bc

Distribution – A species unique to the Cayman Islands. Present throughout dry areas on all three islands, but especially abundant in the dry shrublands at the east ends of each.

Threats – Large-scale loss of habitat due to deforestation is expected to cause massive reductions in population over the generation time for this long-lived species. This trend will eventually overwhelm the vigorous reseeding currently seen in disturbed areas.

Needs – Habitat protection.

Ecology – The Cayman Islands' national tree, with a cultural significance rooted in rope, roofing, and thatch ware, the Silver Thatch is an extremely familiar part of the Cayman Islands' natural and cultural landscape.

It is difficult to accept that such a common, prolific tree can be endangered, but our human perspective is very short term compared to the lifespan of a Silver Thatch which may easily be well over a century. By the time this year's Silver Thatch seedlings are dying of old age, the native forests of the Cayman Islands are likely to have dwindled to whatever areas we have managed to set aside for protection.

Coccothrinax proctorii

Scolosanthus roulstonii

Scolosanthus roulstonii
Proctor

Status – Endangered
EN C1+2a(ii)

Distribution – A species unique to Grand Cayman. Almost exclusively found in the xerophytic shrublands of the east interior. A very small outlying population occurs in the rocky western coast of North Sound.

Threats – The world population of this species is restricted mainly to an arc of dry shrubland at the East End of Grand Cayman, the vast majority of which is unprotected. Just a single large real estate development in this area could cause a drastic reduction in the population of this endemic shrub.

Needs – Habitat protection.

Ecology – *Scolosanthus roulstonii* is a low, wiry shrub growing to 1 m tall, with very small leaves. It grows on karst "cliffrock" formation, and is locally quite common. It bears pale yellow flowers, and white fruits

about 3 mm long. The species was not fully identified and described until 2006.

Along with Corato (*Agave caymanensis*), *Phyllanthus caymanensis* and others, *Scolosanthus roulstonii* is one of a suite of uniquely Caymanian plants and animals which characterise the threatened xerophytic shrubland ecosystem of Grand Cayman.

Scolosanthus roulstonii flowers and fruit

Banara caymanensis leaves

Banara caymanensis
Proctor

**Status – Critically Endangered
CR C2a(i,ii)**

Banara caymanensis

Distribution – A species unique to Cayman Brac and Little Cayman. Occurs in dry shrubland, restricted to two south cliff-top locations on Cayman Brac, and the north-western margin of the Central Forest on Little Cayman.

Threats – An extremely small population places this species at high risk, vulnerable to even small deforestation incidents. Current estimates suggest there may be between 30 and 150 individuals in the wild, in four extremely small clusters.

Needs – Research (reproduction, distribution), *ex situ* propagation, seed banking, awareness, legal protection, and habitat protection.

Ecology – This is a small, woody shrub with wiry, slightly zigzag branches and shiny, minutely toothed leaves. It flowers in June. No sighting of the fruit has ever been reported.

Casearia staffordiae fruits

Casearia staffordiae
Proctor

Status – Critically Endangered
CR B1ab(i,ii,iii,iv,v); C2a(ii)

Distribution – Unique to Grand Cayman. Currently known only from the Mastic forest region.

Threats – The species appears to be restricted at most to the 19.5 km² of dry forest remaining in Grand Cayman, which continues to decline in area. Its population probably numbers less than 250 mature individuals.

Needs – *Ex situ* propagation, seed banking, awareness, legal protection, and habitat protection.

Ecology – This is a distinctive shrub, up to 2 m tall, with long, drooping branches and small leaves. Although first noted in the early 1990s, it remained unidentified until flowers and fruits were finally collected in 2004 and 2005. The fact that such a distinctive plant remained unidentified until so very recently is a reminder that our knowledge of the Cayman Islands' unique flora and fauna is far from complete.

Casearia staffordiae

Allophylus cominia var. *caymanensis*

Allophylus cominia
(L.) Sw.
var. *caymanensis* Proctor

Status – Near Threatened
NT A3bc+4c

Distribution – A variety unique to the three Cayman Islands. Common in dry forests. The species is native to the West Indies and Central America.

Threats – The accelerating deforestation of all three of the Cayman Islands is likely to move this species into a threatened category in a matter of decades.

Needs – Propagation for landscaping, and habitat protection.

Ecology – *Allophylus cominia* var. *caymanensis* produces cream-coloured flower spikes which develop bright red berries. In the wild, it normally grows as an understory shrub, in dappled shade, but it responds well to sun and watering in a cultivated setting. This is an attractive tree-like shrub, which could easily be grown for garden use.

Allophylus cominia var. *caymanensis* fruit

57

Agalinis kingsii

This occasionally spreads from neighbouring agricultural land and burns back stands of *Cladium jamaicense* sedge during the dry season. In the Central Mangrove Wetland, *Agalinis kingsii* colonises exposed peat in trail clearings through Buttonwood shrubland, suggesting that populations might expand widely in the Central Mangrove after widespread tree fall resulting from major hurricanes.

This species is reported to be hemi-parasitic, tapping into root connections to other plants to supplement its own nutrition. An unknown species of ant colonises the raised peat mounds where the *Agalinis* grows. These ants may well be involved in the pollination of *Agalinis'* flowers, and / or the dispersal of its seeds.

REFERENCES: [8]Diochon, A., Burton, F.J., and Garbary, D.J. (2003). Status and ecology of *Agalinis kingsii* (Scrophulariaceae), a rare endemic to the Cayman Islands (Caribbean Sea). *Rhodora* 105 (922): 178–188.

Agalinis kingsii Proctor

Status – Critically Endangered
CR B1ab(v)c(iii,iv)+2ab(v)c(iii,iv)

Distribution – A species unique to Grand Cayman. Occurs in the southern margins of the Salina Reserve sedge wetlands, and in scattered localities within the Central Mangrove Wetland.

Threats – This is an annual herb which appears to undergo extreme population fluctuations. It is restricted geographically and by habitat to a very small area.

Needs – Research (reproduction, ecology), *ex situ* propagation, seed banking, awareness, legal protection, and habitat protection.

Ecology – In the Salina sedge wetlands, *Agalinis kingsii* appears to benefit from fire.[8]

Agalinis kingsii

Turnera triglandulosa
Millsp.

Status – Data Deficient

Distribution – A species unique to Little Cayman and Cayman Brac. Recorded from disturbed ground near coastal roads.

Threats – This species is probably threatened due to its restricted distribution, but no assessment of its abundance, population trend and natural habitat associations has yet been made.

Needs – Research (status and distribution), *ex situ* propagation, seed banking, and habitat protection.

Ecology – Similar at first glance to its close relative, the Cat Bush (*Turnera ulmifolia*), Little Cayman's endemic species has longer, narrower leaves. It often has three glands at the base of the leaf blade, rather than two.

Turnera triglandulosa

Turnera triglandulosa grows as an upright shrub, with straight, slender stems reaching 1.5 m tall. It is easy to cultivate from seed and has been grown successfully in the Queen Elizabeth II Botanic Park on Grand Cayman.

Turnera triglandulosa

Aegiphila caymanensis
Moldenke

Status – Critically Endangered
CR A2ab+3b+4ab; B1ab(i,ii,iii,iv,v)
c(iii)+2ab(i,ii,iii,iv,v)c(iii); C1+2a(i,ii); D

Distribution – A species unique to Grand
Cayman. In recent years it is only known as
a single specimen in the Spotts area and
from a collection in the east interior forest.

Aegiphila caymanensis

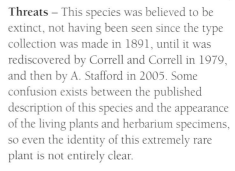

Threats – This species was believed to be
extinct, not having been seen since the type
collection was made in 1891, until it was
rediscovered by Correll and Correll in 1979,
and then by A. Stafford in 2005. Some
confusion exists between the published
description of this species and the appearance
of the living plants and herbarium specimens,
so even the identity of this extremely rare
plant is not entirely clear.

Needs – Research (taxonomy, reproduction,
ecology, distribution), *ex situ* propagation,
seed banking, awareness, legal protection,
and habitat protection.

Ecology – A scrambling shrub or liana
which reaches into the tree canopy, this
species is inconspicuous when not in flower,
and so may be under-reported. However, an
extensive survey of the dry forests and
shrublands of Grand Cayman in the early
1990s yielded no observations of this species
at all.
    Conservation measures for this unique
plant have begun with an attempt at
propagation from cuttings, in the nursery of
the Queen Elizabeth II Botanic Park. The
Spotts specimen was discovered
serendipitously after being blown from the
tree canopy by hurricane Ivan, and is
currently the only known living specimen.

Aegiphila caymanensis

PLANT DESCRIPTIONS
2. Threatened Near-endemic Flora

Fifteen near-endemic taxa, whose distribution includes the Cayman Islands plus one other neighbouring island, are under threat of extinction in the Cayman Islands. One is a variety of Tea Banker (*Pectis caymanensis*) which is described along with its endemic variety on p. 35. The other fourteen follow here.

Tabernaemontana laurifolia L.

Common name – WILD JASMINE

Status in the Cayman Islands – Endangered
EN A3bc+4bc

Distribution – Grand Cayman and Jamaica. In Grand Cayman, it is widespread in dry forests and dry shrublands.

Threats – This species is undergoing significant population declines over decadal time frames as Grand Cayman's forests and shrublands are cleared and converted to human uses.

Needs – Habitat protection.

Ecology – This is a very distinctive understory tree, with glossy, wavy-margined leaves, and twisted yellowish flowers that look like tiny propellers. It is also sometimes called "Slingshot" for the way its branches fork repeatedly. It can be propagated from cuttings: cut stems exude copious white latex. The seeds develop in capsules, which split to show seeds embedded in bright orange pith.

Tabernaemontana laurifolia fruit

Tabernaemontana laurifolia flowers

Lepidaploa divaricata seed

Lepidaploa divaricata Sw.

Common name – CHRISTMAS BLOSSOM

Status in the Cayman Islands – Vulnerable VU A3c+4c

Lepidaploa divaricata flowers

Distribution – All three of the Cayman Islands and Jamaica.

Threats – Like most of the Cayman Islands' native flora, this species is being rapidly lost to ever-expanding man-made landscapes.

Needs – Cultivation and habitat protection.

Ecology – This slightly straggly shrub has conspicuous and attractive purple-blue flowers which bloom towards Christmas each year. As the flowers develop into seeds, the colour is replaced by bright white tufts of seed hairs.

   If cut back periodically, Christmas Blossom has potential as a garden plant. It grows in partial shade on soil in dry shrublands, open woodlands and the edges of clearings. It only thrives where there is at least a little soil, so its distribution in the Cayman Islands is now rather restricted.

Cionosicyos pomiformis fruit

Cionosicyos pomiformis
Griseb.

Common name – Duppy Gourd

Status in the Cayman Islands – Vulnerable VU D2

Distribution – Grand Cayman, in dry forests, and Jamaica.

Threats – The population of this species in the Cayman Islands appears to be restricted to Grand Cayman's 19.5 km² of remaining dry forest, which continues to be eroded by accelerating land clearance.

Needs – Habitat protection.

Ecology – The Duppy Gourd is a high-climbing vine. It is hard to see in the forest tree canopy, and is usually first noticed by its fallen 4 cm-diameter miniature "gourds" lying on the ground. With no obvious origin for these small round apparitions on the forest floor, a Duppy (a West-Indian ghost) might seem the most likely explanation!

Astrocasia tremula

(Griseb.) Webster

**Status in the Cayman Islands – Endangered
EN B2 b(ii,iii,v); C1+2a(ii)**

Distribution – Grand Cayman, in dry forests, and Jamaica.

Threats – In Grand Cayman, this species is relatively uncommon, with a population which may easily number less than 2,500 mature individuals. *Astrocasia* does not occur outside the dry forest ecosystem, and so is directly threatened by the accelerating deforestation of the island.

Needs – Habitat protection.

Ecology – *Astrocasia tremula* is a slow-growing understory tree, usually found growing deep in rocky forests. Male and female flowers, which are born on separate plants, hang on long, delicate stalks, arching from the tree's stems. The green seed capsules are an attractive target for seed-eating birds, including the endemic Cayman Parrot.

Astrocasia tremula

Astrocasia tremula
seed capsules

65

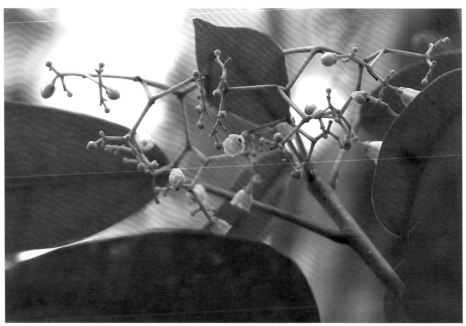

Jatropha divaricata flowers

Jatropha divaricata Sw.

Common name – WILD OIL NUT

Status in the Cayman Islands – Critically Endangered
CR B1ab(v)+2ab(v)

Distribution – Grand Cayman, only in a limited area of the Mastic forest. Also in Jamaica.

Threats – The Cayman Islands population of Wild Oil Nut is restricted to less than 2 km² of dry forest, in the northern and western parts of Grand Cayman's Mastic forest. It is extremely vulnerable to any localised clearance or other impact.

Needs – *Ex situ* propagation, seed banking, awareness, legal protection, and habitat protection.

Ecology – Wild Oil Nut grows in rocky forest on higher land, producing white bell-like flowers and oval seed capsules. It is a small understory tree, which takes readily from cuttings. This species was thought to be unique to Jamaica before it was discovered on Grand Cayman in 1992.

Jatropha divaricata seed capsule

Malpighia cubensis

Malpighia cubensis Kunth

Common name – LADY HAIR

Status in the Cayman Islands – Endangered EN A3bc+4bc

Distribution – Grand Cayman, in dry shrubland. Also in Cuba.

Threats – At risk because the xerophytic shrubland of Grand Cayman's East End, where the majority of Lady Hair survives, is itself at risk from large-scale deforestation.

Needs – Habitat protection.

Ecology – Lady Hair is notorious for the sharp-ended irritant hairs on the undersides of its leaves, which brush off readily onto skin and clothing. They are best removed with sticky tape, or by the edge of a sharp blade.

This is a woody shrub largely restricted to soil zones within rocky shrublands. Lady Hair does have some redeeming features – it bears attractive pink flowers, and red berries which are eaten avidly by birds and rock iguanas.

Malpighia cubensis flowers

Encyclia phoenicia (Lind.) Neum.

Status in the Cayman Islands – Critically Endangered
CR B1ab(i,ii,iii,v)+2ab(i,ii,iii,v); C2a(ii); D

Distribution – Cayman Brac, Little Cayman and Cuba.

Threats – In the Cayman Islands, this orchid has been seen very rarely indeed, and may have been subject to significant collection pressure on Cayman Brac in past years. It is likely to exist as an extremely small, fragmented population at constant threat from deforestation.

Needs – Research (status, distribution), awareness, legal protection, and habitat protection.

Ecology – In Cuba, this is a widespread, variable species. It is now in cultivation commercially and internationally, and is also used in hybridisation. The flowers sometimes smell of vanilla.

Encyclia phoenicia

Encyclia phoenicia

Pleurothallis caymanensis
C.D. Adams

**Status in the Cayman Islands – Vulnerable
VU A3bc+4bc; D2**

Distribution – Grand Cayman, in dry shrublands and transitional forests, and the Peninsula Guanahacabibes of western Cuba.

Threats – This orchid is restricted to low elevation shrubland and forest-shrubland transition. It favours sites downwind from wetlands, and where sinkholes reaching down to groundwater are constantly releasing humidity. Its need for stable humid air restricts the species to a limited area, and a limited number of locations. It is at risk from deforestation throughout its range.

Needs – Habitat protection.

Pleurothallis caymanensis

Ecology – *Pleurothallis caymanensis* usually grows low down on the trunks of trees, in partial shade. In good locations, it can be very abundant, encrusting the trees almost like moss. This is a small species – the leaves are only 6 to 15 mm long, and its flowers are correspondingly tiny.

Pleurothallis caymanensis

*Peperomia
pseudopereskiifolia*

Peperomia pseudopereskiifolia C. DC.

**Status in the Cayman Islands –
Endangered
EN B1ab(ii,iii,v)+2ab(ii,iii,v)**

Distribution – Grand Cayman, Cayman Brac and Cuba.

Threats – This species has a very fragmented distribution in the Cayman Islands' dry forests, with only three locations actually known. Where it occurs, it forms an extensive ground cover, but each location is highly vulnerable to deforestation.

Needs – *Ex situ* propagation, cultivation, awareness, legal protection, and habitat protection.

Ecology – All *Peperomias* are succulent-stemmed with fleshy leaves, and they are quite easy to tell apart by leaf shape – this species has almost tear-drop-shaped leaves, drawn out to a point. Chancing upon rare and localised carpets of *Peperomia* is one of the many pleasures of exploring the Cayman Islands' dry forests, and there are several different species to be found.

Peperomia simplex

Hamilt.

Status in the Cayman Islands – Critically Endangered
CR B1ab(i,ii,iii,v)+2ab(i,ii,iii,v); D

Distribution – Grand Cayman and Jamaica.

Threats – In the Cayman Islands, this species is known only from the Spotts area of Grand Cayman. Here, it was growing over the rock in a fragment of dry forest, which was fast disappearing amid residential development. It may already have been extirpated from this location. Other nearby forest fragments hold two other critically endangered, uniquely Caymanian species, *Pisonia margaretae* and

Peperomia simplex under propagation

Peperomia simplex

Aegiphila caymanensis. These must be the vanishing remnants of a remarkable forest area which has been divided and divided again by roads and subdivisions. It has now been cleared so extensively that its original biodiversity can only be guessed at.

Needs – Research (status and distribution), propagation *ex situ* and for landscaping, awareness, legal protection, and habitat protection.

Ecology – *Peperomia simplex* is a shade-loving, succulent herb, with fleshy spindle-shaped leaves. Like most Peperomias, this species naturally sprouts roots from its stems, and so can be readily propagated from cuttings.

Polygala propinqua flowers

Polygala propinqua

(Britton) Blake

Status in the Cayman Islands – Critically Endangered
CR A3bc+4bc

Distribution – All three Cayman Islands, and Cuba.

Threats – This species is very sparsely distributed in dry shrublands and low forests on Grand Cayman and Little Cayman, and is only slightly more abundant on Cayman Brac. As a slow-growing plant, it is facing very large population losses from deforestation over its next few generations.

Needs – Awareness, legal protection, and habitat protection.

Ecology – This is a small tree, or tree-like shrub, most often found in dry shrubland settings. It produces very small greenish-cream coloured flowers in the leaf axils, which develop into characteristic flattened, heart-shaped capsules. This is a plant which seems to be extremely important to Stripe-headed Tanagers (*Spindalis zena*) on Grand Cayman. These birds return to widely spaced *Polygala* bushes repeatedly to feed on the leaves, which become severely shredded over time. The Tanagers are absent from the Sister Isles, though fossils indicate that they used to be on Cayman Brac at least.

Polygala propinqua

Casearia odorata

Casearia odorata Macf.

Status in the Cayman Islands – Endangered
EN A3bc+4bc; B1ab(i,ii,iii,v)+2ab
(i,ii,iii,v); C1+2a(ii)

Distribution – Grand Cayman and Jamaica.

Threats – This is a species with a small
population in the Cayman Islands, absent
from Cayman Brac and Little Cayman, and
faced with the full force of the deforestation
of Grand Cayman. It is widely spaced in dry
forest settings, and so requires substantial
protected areas to maintain a viable population.

Needs – Research (taxonomy), and habitat
protection.

Ecology – A shrub found rarely in dry forest
understory, this species is also occasionally
abundant in old pastures that are reverting
to forest. It can be recognised as a *Casearia*
by tiny translucent dots in the leaves, visible
when they are held against bright light. It
scrambles up through the shrub layer to a
height of 4–6 m. Its obscure cream-coloured
flowers, clustered in the leaf axils, are
fragrant.

Casearia odorata fruit and flowers

73

Jacquinia proctorii

Stearn

Common name – WASH WOOD

Status in the Cayman Islands – Critically Endangered
CR A3bc+4bc

Distribution – All three of the Cayman Islands, and Jamaica.

Jacquinia proctorii fruit

Jacquinia proctorii

Threats – In the Cayman Islands, Wash Wood is totally dependent on dry forest and shrubland habitats, which are being deforested on time-scales similar to the generation time for this slow-growing species.

Needs – Cultivation, awareness, legal protection, and habitat protection.

Ecology – This species grows as an attractively formed shrub or a small tree, with an even crown, seasonally producing pale yellow flowers and yellowish berries. It often takes forms suggestive of a natural 'bonsai', giving the impression of great age.

In theory at least, this species would make a striking addition to any garden, but little is known about its propagation – especially the extent to which its slow growth in the wild may be speeded up by rich soil and occasional watering.

Daphnopsis occidentalis

(Sw.) Krug & Urban

Common name – BURN NOSE

Status in the Cayman Islands – Endangered EN A4c

Daphnopsis occidentalis fruit

Distribution – All three of the Cayman Islands, and Jamaica.

Threats – Restricted to areas with soil, this species had been hard hit by selective clearing for agriculture, which is now being followed by residential real estate development.

Needs – Habitat protection.

Ecology – A tree-like shrub, or occasionally a small understory tree, Burn Nose is sparsely distributed in forests and shrublands on soil, sometimes persisting in old second-growth areas.

It has a close relative, *Daphnopsis americana*, which carries the same common name: that species grows into a tall canopy tree, also in Grand Cayman's scarce remnants of forest on soil. Both species are most easily recognised by their distinctive, clustered, yellowish-white flower heads.

Daphnopsis occidentalis

Agave caymanensis flowering in the xerophytic shrubland of eastern Little Cayman

RED LIST TABLES

In the following tables, the native plant species and varieties are listed alphabetically, grouped by family.

DISTRIBUTION GROUPS:

GC endemic............. unique to Grand Cayman
LC endemic   ....unique to Little Cayman
CB endemic   ....unique to Cayman Brac
sister isles endemic... unique to Cayman Brac & Little Cayman
caymans endemic...... unique to the three Cayman Islands
near-endemic...........native to the Caymans and one other island
local regional...........native to the NW Caribbean region
greater antillean....... native to the Greater Antilles
west indian.............native to the West Indian region
neotropical............. native to the American tropics
hemispherical..........widespread in the western hemisphere
pantropical............. in the tropics worldwide

Colour is used in the tables to highlight the taxa with the most restricted distributions: all Cayman Islands endemics are in red, near-endemics are in orange, and local regionals are in green.

GC = Grand Cayman
LC = Little Cayman
CB = Cayman Brac

A tick mark ✔ indicates the presence or absence of a plant on each island.

For a brief explanation of the Red List status abbreviations and criteria codes see page 30, above. A full definition of the statuses, criteria and their codes is published at **www.redlist.org**

Family	Species/variety	Common name
AGAVACEAE		
	Agave caymanensis	CORATO
AIZOACEAE		
	Sesuvium maritimum	
	Sesuvium microphyllum	
	Sesuvium portulacastrum	SEA PUSLEY
ALISMATACEAE		
	Sagittaria lancifolia	
AMARANTHACEAE		
	Blutaparon vermiculare	
	Lithophila muscoides	
AMARYLLIDACEAE		
	Hymenocallis latifolia	SPIDER LILY
ANACARDIACEAE		
	Comocladia dentata	MAIDEN PLUM
	Metopium toxiferum	POISON TREE
APOCYNACEAE		
	Echites umbellata	WHITE NIGHTSHADE
	Plumeria obtusa	JASMINE
	Rauvolfia nitida	
	Rhabdadenia biflora	
	Tabernaemontana laurifolia	WILD JASMINE
	Pentalinon luteum	YELLOW NIGHTSHADE
APODANTHACEAE		
	Pilostyles globosa var. *caymanensis*	
ARACEAE		
	Philodendron hederaceum	
ARALIACEAE		
	Dendropanax arboreus	GALIPEE
ASCLEPIDACEAE		
	Metastelma palustre	
	Metastelma picardae	
	Sarcostemma clausum	
ASTERACEAE		
	Ageratum littorale	
	Ambrosia hispida	GERANIUM
	Aster subulatus var. *cubensis*	
	Baccharis dioica	
	Borrichia arborescens	BAY CANDLEWOOD
	Koanophyllon villosum	
	Isocarpha oppositifolia	
	Iva cheiranthifolia	
	Iva imbricata	
	Melanthera aspera	SOFT LEAF

Distribution group	Presence in			National Red List Status
	GC	LC	CB	
caymans endemic	✓	✓	✓	VU A2c+3c+4
neotropical/northern	✓			LC
west indian		✓	✓	LC
pantropical	✓	✓	✓	LC
neotropical	✓			DD, suspected to be at risk
hemispherical	✓	✓	✓	LC
west indian	✓			DD
greater antillean	✓	✓	✓	LC
local regional	✓	✓	✓	LC
greater antillean		✓		EN A3bc+4bc
neotropical	✓	✓	✓	LC
greater antillean	✓	✓	✓	EN A3bc+4bc
greater antillean	✓			CR A4; B1ab(ii,iii,v)+2ab(ii,iii,v)
neotropical	✓	✓		LC
near-endemic	✓			EN A3bc+4bc
west indian	✓			LC
caymans endemic	✓	✓	✓	LC
neotropical	✓			DD
neotropical	✓			CR A3bc+4; B2ab(i,ii,iii,v)
neotropical/northern	✓			DD
near-endemic		✓	✓	DD
neotropical	✓	✓	✓	LC
disjunct	✓			DD
neotropical	✓	✓	✓	LC
neotropical	✓			DD, suspected to be at risk
west indian		✓	✓	DD, suspected to be at risk
west indian	✓	✓	✓	LC
west indian	✓	✓	✓	LC
neotropical			✓	DD, suspected to be at risk
local regional	✓			DD, suspected to be at risk
neotropical	✓			CR B1ab(i,ii,iii,iv,v) +2ab(i,ii,iii,iv,v)
greater antillean		✓	✓	LC

Family	Species/variety	Common name
ASTERACEAE (cont)		
	Pectis caymanensis var. *caymanensis*	TEA BANKER
	Pectis caymanensis var. *robusta*	TEA BANKER
	Pectis linifolia	
	Pluchea carolinensis	
	Pluchea odorata	
	Salmea petrobioides	
	Spilanthes urens	
	Verbesina caymanensis	
	Lepidaploa divaricata	CHRISTMAS BLOSSOM
	Sphagneticola trilobata	MARIGOLD
AVICENNIACEAE		
	Avicennia germinans	BLACK MANGROVE
BASELLACEAE		
	Anredera vesicaria	
BATACEAE		
	Batis maritima	PICKLEWEED
BIGNONIACEAE		
	Catalpa longissima	
	Tabebuia heterophylla	WHITEWOOD
BORAGINACEAE		
	Argusia gnaphalodes	LAVENDAR
	Bourreria venosa	PARROT BERRY
	Cordia brownei	
	Cordia gerascanthus	SPANISH ELM
	Cordia globosa humilis	BLACK SAGE
	Cordia laevigata	CLAM CHERRY
	Cordia sebestena var. *caymanensis*	BROAD LEAF
	Ehretia tinifolia	
	Heliotropium curassavicum	
	Heliotropium humifusum	
	Heliotropium ternatum	
	Tournefortia astrotricha var. *astrotricha*	
	Tournefortia astrotricha var. *subglabra*	
	Tournefortia minuta	
	Tournefortia volubilis	AUNT ELIZA BUSH
BROMELIACEAE		
	Hohenbergia caymanensis	OLD GEORGE
	Tillandsia balbisiana	
	Tillandsia bulbosa	
	Tillandsia fasciculata var. *clavispica*	
	Tillandsia festucoides	
	Tillandsia flexuosa	

Distribution group	Presence in			National Red List Status
	GC	LC	CB	
near-endemic	✓	✓	✓	CR A2acd; B1ab(i,ii,iii,iv,v)+2ab(i,ii,iii,iv,v)
GC endemic	✓			CR A2acd; B1ab(i,ii,iii,iv,v)+2ab(i,ii,iii,iv,v)
neotropical			✓	DD, suspected to be at risk
hemispherical	✓			DD
neotropical	✓			DD
local regional	✓	✓	✓	VU A2c+3c+4c
neotropical	✓	✓		LC
CB endemic			✓	CR B1ab(v)+2ab(v)
near-endemic	✓	✓	✓	VU A3c+4c
hemispherical	✓	✓		LC
hemispherical	✓	✓	✓	EN A2abc+3bc+4
neotropical	✓	✓		DD, suspected to be at risk
hemispheric	✓	✓		VU A2abc+3bc+4
local regional	✓			CR A2c+3c+4c; C1+2a(i,ii); D
west indian	✓	✓	✓	EN A3bc+4bc
neotropical	✓	✓	✓	LC
local regional	✓	✓	✓	LC
near-endemic	✓	✓	✓	LC
neotropical	✓	✓	✓	EN A3bc+4bc
neotropical	✓			VU C1+2a(ii)
greater antillean	✓		✓	CR A3bc+4bc
caymans endemic	✓	✓	✓	VU A3bc+4bc
neotropical	✓			CR A2bc+3bc+4; B1ab(i,ii,iii,v)+2ab C1+2a(i,ii); D
neotropical	✓			DD
local regional	✓	✓	✓	DD, suspected to be at risk
neotropical	✓			DD
local regional	✓		✓	VU A3bc+4bc
near-endemic	✓			DD
greater antillean		✓		DD, suspected to be at risk
neotropical	✓	✓	✓	LC
GC endemic	✓			CR A3c+4c; C1+2a(ii)
neotropical	✓	✓	✓	LC
neotropical	✓	✓		VU A3bc+4bc
near-endemic	✓	✓	✓	LC
neotropical			✓	CR B1ab(v)+2ab(v); C1+2a(i,ii); D
neotropical	✓	✓	✓	LC

Family	Species/variety	Common name
BROMELIACEAE (cont)		
	Tillandsia paucifolia	
	Tillandsia recurvata	OLD MAN'S BEARD
	Tillandsia setacea	
	Tillandsia utriculata	WILD PINE
BURSERACEAE		
	Bursera simaruba	RED BIRCH
BUXACEAE		
	Buxus bahamensis	
CACTACEAE		
	Consolea millspaughii var. *caymanensis*	
	Epiphyllum phyllanthus var. *plattsii*	
	Harrisia gracilis	
	Opuntia dillenii	
	Pilosocereus swartzii	DILDO
	Selenicereus boeckmannii	VINE PEAR
	Selenicereus grandiflorus	VINE PEAR
CANELLACEAE		
	Canella winterana	PEPPER CINNAMON
CAPPARACEAE		
	Capparis cynophallophora	HEADACHE BUSH
	Capparis ferruginea	DEVIL HEAD
	Capparis flexuosa	RAW BONES
	Cleome procumbens	CAT'S WHISKERS
CELASTRACEAE		
	Crossopetalum caymanense	
	Crossopetalum rhacoma	SNAKE BERRY
	Elaeodendron xylocarpum var. *attenuatum*	WILD CALABASH
	Gyminda latifolia	
	Maytenus buxifolia	BASTARD CHELAMELLA
	Schaefferia frutescens	
CHENOPODIACEAE		
	Atriplex pentandra	
	Salicornia bigelovii	
	Salicornia virginica	
CHRYSOBALANACEAE		
	Chrysobalanus icaco	COCOPLUM
CLUSIACEAE		
	Clusia flava	BALSAM
	Clusia rosea	BALSAM
COMBRETACEAE		
	Conocarpus erectus	BUTTONWOOD
	Laguncularia racemosa	WHITE MANGROVE
	Terminalia eriostachya var. *margaretiae*	BLACK MASTIC

Distribution group	Presence in			National Red List Status
	GC	LC	CB	
greater antillean		✓	✓	LC
neotropical/southern	✓	✓	✓	LC
neotropical	✓			VU A3bc+4bc
west indian	✓	✓	✓	LC
neotropical	✓	✓	✓	LC
local regional	✓	✓		CR A3bc+4bc
CB endemic			✓	CR C2a(ii)
CB endemic			✓	CR B1ab(i,ii,iii,v)c(i)+2ab(i,ii,iii,v)c(i); D
local regional	?	?	✓	EN A3bc+4bc
neotropical	✓	✓	✓	EN A3bc+4bc
local regional	✓	✓	✓	CR A3bc+4bc
greater antillean			✓	LC
local regional	✓			LC
west indian	✓	✓	✓	CR A3bc+4bc
neotropical	✓	✓	✓	EN A3bc+4bc
local regional	✓		✓	VU A3bc+4bc
neotropical	✓	✓	✓	LC
local regional	✓			DD
caymans endemic	✓	✓	✓	LC
west indian	✓	✓	✓	LC
local regional	✓			CR A3bc+4bc
west indian	✓	✓	✓	CR A3bc+4bc
local regional		✓	✓	CR A3bc+4bc
west indian	✓	✓	✓	CR A3bc+4bc
neotropical		✓		DD, suspected to be at risk
neotropical/northern		✓		EN A3c+4c
neotropical/northern	✓			DD
neotropical	✓	✓	✓	CR A3bc+4bc
neotropical	✓		✓	LC
neotropical	✓		✓	EN A3bc+4bc
neotropical	✓	✓	✓	VU A3c+4
hemispherical	✓	✓	✓	VU A3c+4
GC endemic	✓			CR A3bc+4bc

Family	Species/variety	Common name
COMMELINACEAE		
	Callisia repens	
	Commelina elegans	WATER GRASS
CONVOLVULACEAE		
	Dichondra repens	
	Evolvulus convolvuloides	
	Evolvulus squamosus	CRAB BUSH
	Ipomoea imperati	
	Ipomoea indica var. *acuminata*	
	Ipomoea passifloroides	
	Ipomoea pes-caprae var. *brasiliensis*	
	Ipomoea violacea	MOONFLOWER
	Jacquemontia havanensis	
CRUCIFERAE		
	Cakile lanceolata	
CUCURBITACEAE		
	Cionosicyos pomiformis	DUPPY GOURD
	Fevillea cordifolia	
CYMODOCEACEAE		
	Halodule wrightii	
	Syringodium filiforme	MANATEE GRASS
CYPERACEAE		
	Cladium jamaicense	CUTTING GRASS
	Cyperus brunneus	CUTTING GRASS
	Cyperus elegans	
	Cyperus filiformis	
	Cyperus floridanus	
	Cyperus ligularis	CUTTING GRASS
	Cyperus planifolius	CUTTING GRASS
	Cyperus swartzii	
	Eleocharis cellulosa	
	Eleocharis interstincta	
	Eleocharis minima	
	Eleocharis mutata	
	Fimbristylis castanea	
	Fimbristylis cymosa var. *spathacea*	
	Fimbristylis ferruginea	
	Fimbristylis spadicea	
	Remirea maritima	
	Rhynchospora colorata	
	Scleria lithosperma	
ERYTHROXYLACEAE		
	Erythroxylum areolatum	SMOKE WOOD
	Erythroxylum confusum	SMOKE WOOD

Distribution group	Presence in			National Red List Status
	GC	LC	CB	
neotropical	✓			VU C2a(ii); D2
neotropical	✓		✓	LC
neotropical		✓		DD, suspected to be at risk
neotropical	✓			LC
local regional		✓		EN A3c+4c
pantropical	✓	✓	✓	DD
pantropical	✓	✓	✓	LC
near-endemic	✓			LC
pantropical	✓	✓	✓	LC
pantropical	✓	✓	✓	LC
west indian	✓	✓	✓	LC
neotropical	✓	✓	✓	LC
near-endemic	✓			VU D2
neotropical	✓			DD, suspected to be at risk
west indian	✓	✓		LC
west indian	✓	✓		LC
pantropical	✓			VU A3c+4
west indian	✓	✓	✓	LC
neotropical	✓		✓	LC
west indian	✓		✓	DD
greater antillean	✓	✓	✓	DD
hemispherical	✓	✓	✓	LC
west indian	✓			DD
west indian	✓			DD
neotropical	✓		✓	LC
neotropical	✓			DD
neotropical	✓			DD
hemispheric	✓			LC
neotropical/northern	✓			DD
pantropical	✓	✓	✓	LC
pantropical	✓	✓	✓	LC
neotropical/northern	✓			LC
pantropical	✓	✓		CR C2a(i)
neotropical	✓			LC
pantropical	✓	✓	✓	LC
neotropical	✓	✓	✓	LC
greater antillean	✓			CR B2ab(i,ii,iii,v)

Family	Species/variety	Common name
ERYTHROXYLACEAE (cont)		
	Erythroxylum rotundifolium	RAT WOOD
EUPHORBIACEAE		
	Acalypha chamaedrifolia	
	Adelia ricinella	
	Argythamnia proctorii	CAYMAN SILVERBUSH
	Astrocasia tremula	
	Bernardia dichotoma	
	Chamaesyce blodgettii	
	Chamaesyce bruntii	
	Chamaesyce mesembrianthemifolia	TITTIE MOLLY
	Chamaesyce ophthalmica	
	Chamaesyce torralbasii	
	Chascotheca domingensis	
	Chascotheca neopeltandra	
	Croton linearis	ROSEMARY
	Croton lucidus	
	Croton nitens	WILD CINNAMON
	Croton rosmarinoides	
	Drypetes sp. indet.	
	Euphorbia cassythoides	
	Euphorbia trichotoma	
	Gymnanthes lucida	CRAB BUSH
	Hippomane mancinella	MANCHINEEL
	Jatropha divaricata	WILD OIL NUT
	Margaritaria nobilis	
	Phyllanthus angustifolius	DUPPY BUSH
	Phyllanthus caymanensis	
	Phyllanthus nutans var. grisebachianus	
	Phyllanthus nutans var. nutans	
	Picrodendron baccatum	BITTER PLUM
	Savia erythroxyloides	WILD COCOPLUM
	Securinega acidoton	
	Tragia volubilis	ITCHING VINE
FABACEAE		
	Bauhinia divaricata	BULL HOOF
	Caesalpinia bonduc	COCKSPUR
	Caesalpinia intermedia	YELLOW NICKAR
	Caesalpinia wrightiana	YELLOW NICKAR
	Calliandra cubensis	
	Canavalia nitida	HORSE BEAN
	Canavalia rosea	SEA BEAN
	Centrosema virginianum	

Distribution group	Presence in			National Red List Status
	GC	LC	CB	
greater antillean	✓	✓	✓	EN A3bc+4bc
west indian			✓	DD
west indian	✓	✓	✓	CR A3bc+4bc
caymans endemic	✓	✓	✓	LC
near-endemic	✓			EN B2 b(ii,iii,v); C1+2a(ii)
west indian	✓	✓	✓	LC
neotropical	✓	✓	✓	LC
LC endemic		✓		CR B1b(ii,iii)
neotropical	✓	✓	✓	LC
west indian	✓	✓	✓	LC
near-endemic			✓	DD
local regional	✓		✓	DD, suspected to be at risk
local regional	✓		✓	VU A3bc+4bc
greater antillean	✓	✓	✓	LC
greater antillean	✓	✓	✓	LC
neotropical	✓	✓	✓	EN A3bc+4bc
local regional	✓	✓	✓	EN A3c+4c; B1ab(i,ii,iii,v) +2ab(i,ii,iii,v); C1
unknown	✓			CR B1ab(i,ii,iii,iv,v)c(i,iii)+2ab(i,ii,iii,iv,v) c(i,iii); D
local regional	✓			VU A3bc+4bc
neotropical	✓	✓	✓	DD
west indian	✓	✓	✓	CR A3bc+4bc
neotropical	✓	✓	✓	EN A3bc+4bc
near-endemic	✓			CR B1ab(v)+2ab(v)
neotropical	✓			CR D
local regional	✓	✓	✓	VU A3bc+4bc
caymans endemic	✓	✓	✓	VU D1+2
near-endemic		✓		DD
local regional	✓	✓	✓	VU A3bc+4bc
greater antillean	✓	✓	✓	EN A3bc+4bc
greater antillean	✓	✓	✓	VU A3bc+4bc
west indian	✓			DD
neotropical	✓			DD
neotropical	✓	✓	✓	LC
pantropical	✓	✓	✓	LC
local regional	✓	✓		VU C1
local regional		✓		DD, suspected to be at risk
local regional	✓	✓	✓	EN B1ab(ii,iii,v)+2ab(ii,iii,v); C1+2a(i)
west indian	✓			EN C2a(ii)
pantropical	✓	✓	✓	LC
neotropical	✓		✓	LC

Family	Species/variety	Common name
FABACEAE (cont)		
	Chamaecrista lineata	STORM WEED
	Chamaecrista nictitans var. *aspera*	WILD SHAME FACE
	Dalbergia brownei	COCOON
	Dalbergia ecastaphyllum	
	Erythrina velutina	COCKSPUR TREE
	Piscidia piscipula	DOGWOOD
	Sophora tomentosa	MICAR
	Tephrosia cinerea	
	Tephrosia senna	
GENTIANACEAE		
	Eustoma exaltatum	
	Voyria parasitica	
GOODENIACEAE		
	Scaevola plumieri	BAY BALSAM
HYDROCHARITACEAE		
	Halophila baillonis	
	Halophila engelmannii	
	Thalassia testudinum	TURTLE GRASS
LABIATAE		
	Salvia caymanensis	
	Ocimum micranthum	PIMENTO BASIL
LAURACEAE		
	Cassytha filiformis	DODDER
	Licaria triandra	
	Ocotea coriacea	SWEETWOOD
LORANTHACEAE		
	Dendropemon caymanensis	
LYTHRACEAE		
	Ammania latifolia	
MALPIGHIACEAE		
	Bunchosia media	
	Malpighia cubensis	LADY HAIR
MALVACEAE		
	Bastardia viscosa parvifolia	
	Hibiscus pernambucensis	SEASIDE MAHOE
	Kosteletzkya pentasperma	
	Malvastrum corchorifolium	
	Malvaviscus arboreus var. *cubensis*	MAHOE
	Sida ciliaris	
	Thespesia populnea	PLOPNUT
MELIACEAE		
	Cedrela odorata	CEDAR

Distribution group	Presence in			National Red List Status
	GC	LC	CB	
local regional	✓			DD, suspected to be at risk
neotropical	✓			LC
neotropical	✓			EN A2abc+3bc+4
hemispherical	✓	✓		EN A2abc+3bc+4
neotropical	✓			CR B2ab(ii,iii,v); D
neotropical	✓			EN A3bc+4bc; B1ab(ii,iii,iv,v)+2abC1+2a(ii)
pantropical	✓			CR A2abc
neotropical	✓			LC
neotropical			✓	DD
neotropical	✓			LC
neotropical	✓			DD, suspected to be at risk
hemispherical	✓	✓	✓	CR A1abce; B1ab(i,ii,iii,iv,v)
neotropical	✓			LC
neotropical	✓			LC
neotropical	✓	✓	✓	LC
GC endemic	✓			CR B1ab(i,ii,iii,iv,v)c(ii,iv)+2ab(i,ii,iii,iv,v) c(ii,iv)
neotropical	✓	✓		DD
pantropical	✓	✓	✓	LC
neotropical	✓			CR B1ab(i,ii,iii,v)+2ab(i,ii,iii,v)
west indian	✓		✓	VU A3bc+4bc
LC endemic		✓		CR C2a(i,ii)
neotropical	✓			LC
local regional	✓	✓		VU A3bc+4bc
near-endemic	✓			EN A3bc+4bc
neotropical	✓			DD
pantropical	✓	✓		EN C2a(i)
neotropical	✓			LC
west indian	✓		✓	LC
greater antillean	✓	✓	✓	LC
neotropical	✓		✓	LC
pantropical	✓	✓	✓	EN A3bc+4bc
neotropical	✓		✓	CR A2bcde+3bce+4

Family	Species/variety	Common name
MELIACEAE (cont)		
	Swietenia mahagoni	MAHOGANY
	Trichilia glabra	BASTARD MAHOGANY
	Trichilia havanensis	
MENISPERMACEAE		
	Cissampelos pareira	QUACORI
MORACEAE		
	Ficus aurea	WILD FIG
	Ficus citrifolia	BARREN FIG
	Maclura tinctoria	FUSTIC
MYOPORACEAE		
	Bontia daphnoides	
MYRICACEAE		
	Myrica cerifera	BAYBERRY
MYRSINACEAE		
	Myrsine acrantha	
MYRTACEAE		
	Calyptranthes pallens	BASTARD STRAWBERRY
	Eugenia axillaris	STRAWBERRY
	Eugenia biflora	
	Eugenia foetida	
	Myrcianthes fragrans	CHERRY
NYCTAGINACEAE		
	Boerhavia coccinea	CHICK WEED
	Boerhavia erecta	BROOM WEED
	Guapira discolor	CABBAGE TREE
	Pisonia aculeata	
	Pisonia margaretae	
NYMPHAEACEAE		
	Nymphaea ampla	WATER LILY
OLACACEAE		
	Schoepfia chrysophylloides	
OLEACEAE		
	Chionanthus caymanensis	IRONWOOD
	Forestiera segregata	
ORCHIDACEAE		
	Beloglottis costaricensis	
	Brassavola nodosa	
	Cyclopogon cranichoides	
	Cyclopogon elatus	
	Cyrtopodium punctatum	
	Dendrophylax fawcettii	GHOST ORCHID
	Eltroplectris calcarata	

Distribution group	Presence in			National Red List Status
	GC	**LC**	**CB**	
west indian	✓	✓	✓	EN A2bcd+3bc+4
local regional	✓	✓	✓	EN A3bc+4bc
neotropical	✓			CR A2abc+3c+4;B1ab(i,ii,iii,v)+2ab(i,ii,iii,v); C1+2a(ii)
pantropical	✓		✓	VU C1
greater antillean	✓	✓	✓	VU A3bc+4bc
neotropical	✓	✓	✓	EN A3bc+4bc
neotropical	✓		✓	CR A3bc+4bc
neotropical	✓	✓	✓	DD
neotropical	✓			VU A3c+4c
local regional	✓			CR A3bc+4bc
west indian	✓	✓	✓	EN A3bc+4bc
neotropical	✓	✓	✓	LC
neotropical	✓			DD, suspected to be at risk
neotropical			✓	DD, suspected to be at risk
neotropical	✓	✓	✓	EN A3bc+4bc
neotropical	✓		✓	LC
neotropical	✓	✓	✓	LC
greater antillean	✓	✓	✓	LC
pantropical		✓	✓	EN B1ab(ii,iii,v)+2ab(ii,iii,v)
GC endemic	✓			CR C1+2a(i,ii); D
neotropical	✓			DD
local regional	✓	✓	✓	EN A3bc+4bc
caymans endemic	✓	✓	✓	EN A3bc+4bc
west indian	✓	✓		EN B1ab(ii,iii,v)+2ab(ii,iii,v)
neotropical	✓			CR B1ab(i,ii,iii,v)+2ab(i,ii,iii,v); C2a(ii); D
neotropical	✓			CR B1ab(i,ii,iii,v)+2ab(i,ii,iii,v); C2a(ii); D
neotropical	✓			CR D
neotropical	✓			EN C2a(ii)
neotropical	✓			CR B1ab(i,ii,iii,v)+2ab(i,ii,iii,v); C2a(i); D
GC endemic	✓			CR A2abcd+4abcd
neotropical	✓		?	CR B2ab(i,ii,iii,iv,v)

Family	Species/variety	Common name
ORCHIDACEAE (cont)		
	Prosthechea boothiana	
	Prosthechea cochleata	
	Encyclia kingsii	
	Encyclia phoenicia	
	Epidendrum nocturnum	
	Epidendrum rigidum	
	Dendrophylax porrectus	
	Ionopsis utricularioides	
	Pleurothallis caymanensis	
	Polystachya concreta	
	Prescottia oligantha	
	Myrmecophila thompsoniana	BANANA ORCHID
	Myrmecophila thompsoniana var. minor	BANANA ORCHID
	Myrmecophila thompsoniana var. thomsoniana	BANANA ORCHID
	Sacoila lanceolata	
	Tolumnia calochila	
	Tolumnia variegata	
	Triphora gentianoides	
	Tropidia polystachya	
	Vanilla claviculata	
PALMAE		
	Coccothrinax proctorii	SILVER THATCH
	Roystonea regia	ROYAL PALM
	Thrinax radiata	BULL THATCH
PAPAVERACEAE		
	Argemone mexicana	THOM THISTLE
PASSIFLORACEAE		
	Passiflora cupraea	
	Passiflora suberosa	WILD PUMPKIN
PHYTOLACCACEAE		
	Rivina humilis	FOWL BERRY
	Trichostigma octandrum	
PIPERACEAE		
	Peperomia glabella	
	Peperomia obtusifolia	VINE BALSAM
	Peperomia pseudopereskiifolia	
	Peperomia simplex	
	Piper amalago	PEPPER ELDER
POACEAE		
	Andropogon glomeratus	
	Cenchrus gracillimus	
	Cenchrus incertus	
	Cenchrus tribuloides	

Distribution group	Presence in			National Red List Status
	GC	LC	CB	
neotropical	✓			Near threatened A3c+4c
west indian	✓		✓	CR B1ab(i,ii,iii,v)+2ab(i,ii,iii,v); C2a(ii); D
sister isles endemic		✓	✓	CR A2abcd+3bc+4bc; C1+2a(i,ii); D
near-endemic			✓	CR B1ab(i,ii,iii,v)+2ab(i,ii,iii,v); C2a(ii); D
neotropical	✓			CR B1ab(i,ii,iii,v)+2ab(i,ii,iii,v); C2a(ii); D
neotropical	✓			CR B1ab(i,ii,iii,v)+2ab(i,ii,iii,v); C2a(ii); D
neotropical	✓			CR B1ab(i,ii,iii,v)+2ab(i,ii,iii,v); C2a(ii); D
neotropical	✓			CR B1ab(i,ii,iii,v)+2ab(i,ii,iii,v); C2a(ii); D
near-endemic	✓			VU A3bc+4bc; D2
hemispherical		✓		CR D
neotropical	✓			VU D2
caymans endemic	✓	✓	✓	EN A3bc+4bc
sister isles endemic		✓	✓	EN A3bc+4bc
GC endemic	✓			EN A3bc+4bc
neotropical	✓			DD
greater antillean	✓			CR B1ab(i,ii,iii,v)+2ab(i,ii,iii,v); C2a(ii); D
greater antillean	✓			CR A2abcd+3cd+4cd
neotropical	✓			CR D
neotropical	✓			CR D
greater antillean	✓			VU B2ab(i,ii,iii,iv,v)
caymans endemic	✓	✓	✓	EN A3bc+4bc
local regional	✓			EN A3bc+4bc; C1+2a(ii)
greater antillean	✓	✓		CR A3bc+4bc
neotropical	✓		✓	LC
local regional	✓	✓	✓	LC
hemispherical	✓	✓	✓	LC
neotropical	✓	✓	✓	LC
neotropical			✓	DD
neotropical	✓			CR A3bc+4bc; B2ab(i,ii,iii,iv,v)
neotropical	✓			EN A3bc+4bc; B1ab(i,ii,iii,v)+2ab(i,ii,iii,v)
near-endemic	✓		✓	EN B1ab(ii,iii,v)+2ab(ii,iii,v)
near-endemic	✓			CR B1ab(i,ii,iii,v)+2ab(i,ii,iii,v); D
neotropical	✓			DD, suspected to be at risk
neotropical	✓			LC
greater antillean	✓			DD
neotropical	✓	✓	✓	LC
neotropical	✓	✓	✓	LC

Family	Species/variety	Common name
POACEAE (cont)		
	Chloris petraea	
	Distichlis spicata	NO MAN'S CONQUOR
	Echinochloa walteri	
	Eragrostis domingensis	
	Lasiacis divaricata	
	Leptochloa fascicularis	
	Paspalum blodgettii	
	Paspalum distichum	
	Paspalum vaginatum	
	Pharus glaber	
	Spartina patens	
	Sporobolus domingensis	
	Sporobolus virginicus	NO MAN'S CONQUOR
POLYGALACEAE		
	Polygala propinqua	
POLYGONACEAE		
	Coccoloba uvifera	SEA GRAPE
	Polygonum densiflorum	
	Polygonum punctatum	
POLYPODIACEAE		
	Acrostichum aureum	
	Acrostichum danaeifolium	
	Adiantum melanoleucum	
	Adiantum tenerum	
	Blechnum serrulatum	
	Cheilanthes microphylla	
	Nephrolepis biserrata	
	Nephrolepis exaltata	
	Polypodium aureum	
	Polypodium dispersum	
	Polypodium heterophyllum	
	Polypodium phyllitidis	COW TONGUE
	Polypodium polypodioides	RESURRECTION FERN
	Pteris longifolia var. bahamensis	
	Tectaria incisa	
	Thelypteris augescens	
	Thelypteris interrupta	
	Thelypteris kunthii	
	Thelypteris reptans	
PORTULACACEAE		
	Portulaca halimoides	
	Portulaca pilosa	TEN-O'CLOCK
	Portulaca rubricaulis	

Distribution group	Presence in			National Red List Status
	GC	LC	CB	
neotropical	✓	✓	✓	LC
neotropical/northern	✓			LC
greater antillean	✓			DD
west indian	✓	✓	✓	LC
neotropical	✓	✓	✓	LC
neotropical/northern	✓			LC
neotropical	✓	✓	✓	LC
hemispherical	✓			LC
pantropical	✓	✓	✓	LC
neotropical	✓			DD
west indian	✓	✓		LC
greater antillean	✓			LC
neotropical	✓	✓	✓	LC
near-endemic	✓	✓	✓	CR A3bc+4bc
neotropical	✓	✓	✓	CR A3bc+4bc
pantropical	✓			DD
hemispherical	✓			DD
pantropical	✓	✓	✓	LC
hemispherical	✓	✓		DD, suspected to be at risk
greater antillean	✓			EN D
neotropical	✓			EN D
neotropical	✓			VU B2ab(i,ii,iii,iv,v)
neotropical	✓			DD
pantropical	✓			LC
greater antillean	✓		✓	LC
neotropical	✓			CR C1+2a(i,ii); D
neotropical	✓			Near threatened A3bc+4bc
west indian	✓			EN B2ab(i,ii,iii,iv,v); D
neotropical	✓			Near threatened A3bc+4bc
hemispherical	✓	✓	✓	Near threatened A3bc+4bc
greater antillean	✓		✓	DD
neotropical	✓			DD
greater antillean	✓			EN D
pantropical	✓			VU D1+2
neotropical	✓		✓	LC
neotropical	✓			LC
west indian	✓			DD
neotropical	✓		✓	DD
neotropical	✓	✓	✓	LC

Family	Species/variety	Common name
PORTULACACEAE (cont)		
	Portulaca tuberculata	
PSILOTACEAE		
	Psilotum nudum	
RHAMNACEAE		
	Colubrina arborescens	RED HEART
	Colubrina cubensis	CAJON
	Colubrina elliptica	WILD GUAVA
RHIZOPHORACEAE		
	Rhizophora mangle	RED MANGROVE
RUBIACEAE		
	Antirhea lucida	
	Catesbaea parviflora	
	Chiococca alba	SNOW BERRY
	Chiococca parvifolia	
	Erithalis fruticosa	BLACK CANDLEWOOD
	Ernodea littoralis	GUANA BERRY
	Exostema caribaeum	BASTARD IRONWOOD
	Faramea occidentalis	
	Guettarda elliptica	
	Hamelia cuprea	
	Morinda royoc	YELLOW ROOT
	Psychotria nervosa	STRONG BACK
	Randia aculeata	LANCE WOOD
	Rhachicallis americana	SANDFLY BUSH
	Scolosanthus roulstonii	
	Spermacoce tetraquetra	
	Strumpfia maritima	
RUPPIACEAE		
	Ruppia cirrhosa	
	Ruppia maritima	
RUTACEAE		
	Amyris elemifera	CANDLEWOOD
	Zanthoxylum coriaceum	SHAKE HAND
	Zanthoxylum flavum	SATINWOOD
SALICACEAE		
	Banara caymanensis	
	Casearia aculeata	THORN PRICKLE
	Casearia guianensis	WILD COFFEE
	Casearia hirsuta	WILD COFFEE
	Casearia odorata	
	Casearia staffordiae	
	Casearia sylvestris	

Distribution group	Presence in			National Red List Status
	GC	LC	CB	
local regional		✓	✓	DD
pantropical	✓			DD
neotropical	✓			CR A3c+4c; B2a+b(i,ii,iii,iv,v)
greater antillean	✓	✓	✓	LC
neotropical	✓	✓	✓	EN A3bc+4bc
hemispherical	✓	✓	✓	Near threatened A3bc+4bc
greater antillean	✓	✓		CR A3bc+4bc
local regional	✓	✓	✓	VU A3bc+4bc
neotropical	✓	✓	✓	LC
west indian		✓	✓	VU C1; D2
neotropical	✓	✓	✓	EN A3bc+4bc
west indian	✓	✓	✓	LC
neotropical	✓	✓	✓	EN A3bc+4bc
neotropical	✓			CR B1ab(i,ii,iii,v)+2ab(i,ii,iii,v)
neotropical	✓	✓	✓	EN A3bc+4bc
greater antillean	✓	✓	✓	VU A3bc+4bc
neotropical	✓			LC
neotropical	✓			VU A3bc+4bc
neotropical	✓	✓	✓	LC
local regional	✓	✓	✓	LC
GC endemic	✓			EN C1+2a(ii)
local regional			✓	DD
west indian	✓	✓	✓	LC
cosmopolitan	✓			LC
cosmopolitan	✓		✓	LC
neotropical	✓	✓	✓	EN A3bc+4bc
greater antillean	✓	✓	✓	CR A3bc+4bc
west indian	✓	✓	✓	CR A3bc+4bc
sister Isles endemic		✓	✓	CR C2a(i,ii)
neotropics	✓			VU A3bc+4bc
neotropical			✓	CR B1ab(i,ii,iii,v)
neptropical	✓		✓	VU A3bc+4bc
near-endemic	✓		✓	EN A3bc+4bc; B1ab(i,ii,iii,v)+2ab (i,ii,iii,v); C1+2a(ii)
GC endemic	✓			CR B1ab(i,ii,iii,iv,v); C2a(ii)
neotropical	✓			CR B1ab(i,ii,iii,v)

Family	Species/variety	Common name
SALICACEAE (cont)		
	Xylosma bahamense	SHAKE HAND
	Zuelania guidonia	JEREMIAH BUSH
SAPINDACEAE		
	Allophylus cominia var. *caymanensis*	
	Cardiospermum corindum	
	Cardiospermum halicacabum	
	Dodonaea viscosa	
	Exothea paniculata	WILD GINEP
	Hypelate trifoliata	PLUMPERRA
SAPOTACEAE		
	Sideroxylon foetidissimum	MASTIC
	Sideroxylon horridum	GREEN THORN
	Sideroxylon salicifolium	WILD SAPODILLA
SCROPHULARIACEAE		
	Agalinis kingsii	
	Scoparia dulcis	
	Stemodia maritima	
SIMAROUBACEAE		
	Alvaradoa amorphoides	WILD SPANISH ARMADA
SMILACACEAE		
	Smilax havanensis	WIRE WISS
SOLANACEAE		
	Cestrum diurnum var. *marcianum*	
	Cestrum diurnum var. *venenatum*	
	Solandra longiflora	
	Solanum bahamense	
	Solanum havanense	
	Solanum lanceifolium	
STERCULIACEAE		
	Helicteres jamaicensis	SCREW BUSH
	Melochia tomentosa	VELVET LEAF
	Neoregnellia cubensis	
SURIANACEAE		
	Suriana maritima	JUNIPER
THEOPHRASTACEAE		
	Jacquinia keyensis	WASH WOOD
	Jacquinia proctorii	
THYMELACEAE		
	Daphnopsis americana	
	Daphnopsis occidentalis	BURN NOSE
TILIACEAE		
	Corchorus hirsutus	

Distribution group	Presence in			National Red List Status
	GC	**LC**	**CB**	
local regional	✓			EN A3bc+4bc
neotropical	✓	✓	✓	CR A3bc+4bc
caymans endemic	✓	✓	✓	Near threatened A3bc+4c
pantropical	✓	✓	✓	LC
pantropical	✓		✓	DD
pantropical	✓	✓		DD, suspected to be at risk
neotropical			✓	EN A3bc+4bc
west indian	✓	✓	✓	EN A3bc+4bc
west indian	✓		✓	CR A3bc+4bc
local regional	✓	✓	✓	EN A3bc+4bc
neotropical	✓	✓	✓	EN A3bc+4bc
GC endemic	✓			CR B1ab(v)c(iii,iv)+2ab(v)c(iii,iv)
pantropical	✓			DD
neotropical	✓			LC
neotropical	✓	✓		EN A3bc+4bc
greater antillean	✓			LC
near-endemic	✓			DD, suspected to be at risk
local regional	✓		✓	VU C1
local regional			✓	VU D2
greater antillean	✓	✓	✓	LC
greater antillean	✓			DD
west indian	✓			DD
neotropical	✓	✓	✓	VU A3bc+4bc
neotropical	✓	✓	✓	LC
local regional	✓			CR A2ac+4abc; B1ab(i,ii,iii,v)+2ab(i,ii,iii,v); C1+2a(ii); D
pantropical	✓	✓	✓	LC
greater antillean	✓		✓	EN B1ab(i,ii,iii,iv,v)+2ab(i,ii,iii,iv,v)+C2a(i)
near-endemic	✓	✓	✓	CR A3bc+4bc
neotropical	✓			CR B2ab(ii,iii,iv,v)
near-endemic	✓	✓	✓	EN A4c
hemispherical	✓			LC

Family	Species/variety	Common name
TURNERACEAE		
	Turnera diffusa	
	Turnera triglandulosa	
	Turnera ulmifolia	CAT BUSH
TYPHACEAE		
	Typha domingensis	CAT TAIL
ULMACEAE		
	Celtis iguanaea	
	Celtis trinervia	
	Trema lamarckianum	
UMBELLIFERAE		
	Centella asiatica	
URTICACEAE		
	Pilea herniarioides	
	Pilea microphylla var. *succulenta*	
VERBENACEAE		
	Aegiphila caymanensis	
	Aegiphila elata	
	Citharexylum fruticosum	YELLOW FIDDLEWOOD
	Clerodendrum aculeatum var. *aculeatum*	CAT CLAW
	Clerodendrum aculeatum var. *gracile*	CAT CLAW
	Duranta erecta	
	Lantana aculeata	
	Lantana bahamensis	
	Lantana camara	WHITE SAGE
	Lantana involucrata	BITTER SAGE
	Lantana urticifolia	SWEET SAGE
	Petitia domingensis	FIDDLEWOOD
	Lippia alba	
	Lippia nodiflora	MATCH HEAD
	Stachytarpheta jamaicensis	VERVINE
VISCACEAE		
	Phoradendron quadrangulare	SCORN-THE-GROUND
	Phoradendron rubrum	SCORN-THE-GROUND
	Phoradendron trinervium	
VITACEAE		
	Cissus microcarpa	PUDDING WITHE
	Cissus trifoliata	
	Cissus verticillata	
ZAMIACEAE		
	Zamia integrifolia	BULRUSH
ZYGOPHYLLACEAE		
	Tribulus cistoides	JIM CARTER WEED

Distribution group	Presence in			National Red List Status
	GC	LC	CB	
neotropical		✓		DD
sister isles endemic		✓		DD, suspected to be at risk
neotropical	✓	✓	✓	LC
neotropical/southern	✓			VU A2abc+3bc
neotropical	✓	✓	✓	EN A3bc+4bc
greater antillean	✓			CR A3bc+4bc
west indian	✓	✓		VU A1abc+2bc
pantropical	✓			DD
west indian	✓			DD
local regional	✓		✓	DD
GC endemic	✓			CR A2ab+3b+4ab; B1ab(i,ii,iii,iv,v)c(iii)+2ab(i,ii,iii,iv,v)c(iii); C1+2a(i,ii); D
neotropical	✓			DD, suspected to be at risk
neotropical	✓	✓	✓	EN A3bc+4bc
neotropical	✓	✓	✓	DD
near-endemic	✓			DD, suspected to be at risk
neotropical	✓			VU C1
hemispherical		✓	✓	Data deficient
local regional	✓			DD
neotropical	✓	✓	✓	LC
neotropical	✓	✓	✓	LC
west indian	✓			DD
greater antillean	✓		✓	EN A3bc+4bc
neotropical/southern	✓		✓	DD
pantropical	✓	✓	✓	LC
neotropical	✓	✓	✓	LC
neotropical	✓	✓	✓	LC
local regional	✓	✓	✓	DD
neotropical	✓	✓		DD
neotropical	✓	✓	✓	LC
neotropical	✓	✓	✓	LC
neotropical	✓	✓	✓	VU C1+2a(ii); D2
greater antillean	✓	✓	✓	CR A3bc+4bc
neotropical	✓	✓	✓	LC

BIBLIOGRAPHY

Brunt, M.A. (1984). Environment and plant communities. In: *Flora of the Cayman Islands*, G.R. Proctor, pp. 5–58, *Kew Bulletin Additional Series* X1. The Royal Botanic Gardens, Kew.

Burton, F.J. (1994). Climate and tides of the Cayman Islands. In: *The Cayman Islands: Natural History and Biogeography*, ed. M.A. Brunt & J.E. Davies, pp. 51–60. Kluwer Academic Publishers, Dordrecht, The Netherlands.

Diochon, A., Burton, F.J., and Garbary, D.J. (2003). Status and ecology of *Agalinis kingsii* (Scrophulariaceae), a rare endemic to the Cayman Islands (Caribbean Sea). *Rhodora* 105 (922): 178–188.

Jones, B., Hunter, I.G. and Kyser, K. (1994). Stratigraphy of the Bluff Formation (Miocene-Pliocene) and the Newly Defined Brac Formation (Oligocene), Cayman Brac, British West Indies . Caribbean Journal of Science, Vol. 30, No. 1-2, 30-51, 1994

Jones, B. and Hunter, I.G. (1990). Pleistocene paleogeography and sea levels on the Cayman Islands, British West Indies. Coral Reefs 9:81-91.

National Trust for the Cayman Islands Law 22 of 1987 (1997 Revision).

Proctor, G.R. (in press). Flora of the Cayman Islands, second edition. The Royal Botanic Gardens, Kew.

INTERNET RESOURCES

www.redlist.org

www.iucn.org

INDEX OF SCIENTIFIC NAMES

Page numbers indexed in bold type are the main description pages for the endemic and near-endemic species and varieties.

A

Aegiphila caymanensis **60**, 71
Agalinis kingsii **58**
Agavaceae 33
Agave caymanensis 16, **33**, 54
Allophylus cominia **57**
   var. *caymanensis* **57**
Apocynaceae 62
Apodanthaceae 34
Argythamnia proctorii **43**
Asteraceae 35, 36, 63
Astrocasia tremula **65**
Avicennia germinans 15

B

Banara caymanensis **55**
Boraginaceae 37
Bromeliaceae 38
Bryophyllum pinnatum 36
Bauhinia divaricata 34
Bursera simaruba 15

C

Cactaceae 39, 40
Capparis cynophallophora 47
Casearia odorata **73**
Casearia staffordiae **56**
Casuarina equisetifolia 21, 37
Celastraceae 41
Chamaesyce bruntii **44**
Chamaesyce prostrata 44

Chionanthus caymanensis

Chionanthus caymanensis 20, **49**
Cionosicyos pomiformis **64**
Cladium jamaicense 28, 58
Coccothrinax proctorii **53**
Coccoloba uvifera 16
Combretaceae 42
Conocarpus erectus 15
Consolea millspaughii **39**
   var. *caymanensis* **39**
Consolea moniliformis 39
Cordia sebestena **37**
   var. *caymanensis* **37**
Crossopetalum caymanense **41**
Cucurbitaceae 64

D

Daphnopsis americana 75
Daphnopsis occidentalis **75**
Dendropemon caymanensis **47**
Dendrophylax fawcettii **50**

E

Encyclia kingsii **51**
Encyclia phoenicia **68**
Epiphyllum phyllanthus **40**
   var. *plattsii* **40**
Erithalis fruticosa 47
Euphorbiaceae 43, 44, 45, 65, 66

G

Guapira discolor 15

H

Haematoxylum campechianum 21
Hohenbergia caymanensis 9, **38**

J

Jacquinia proctorii **74**
Jatropha divaricata **66**

L

Labiatae 46
Laguncularia racemosa 15
Lepidaploa divaricata **63**
Loranthaceae 47

M

Malpighiaceae 67
Malpighia cubensis **67**
Metopium toxiferum 15
Myrmecophila thomsoniana **52**
        var. *minor* **52**
        var. *thomsoniana* **52**

N

Nyctaginaceae 48

O

Oleaceae 49
Orchidaceae 50, 51, 52, 68, 69

P

Palmae 53
Pectis caymanensis **35**, 61
        var. *caymanensis* **35**
        var. *robusta* **35**
Peperomia pseudopereskiifolia **70**
Peperomia simplex **71**
Phyllanthus caymanensis **45**, 54
Pilosocereus swartzii 15

Pilostyles globosa **34**
        var. *caymanensis* **34**
Piperaceae 70, 71
Pisonia margaretae **48**, 71
Pleurothallis caymanensis **69**
Polygalaceae 72
Polygala propinqua **72**

R

Rhizophora mangle 15
Roystonea regia 15
Rubiaceae 54

S

Salicaceae 55, 56, 73
Salvia caymanensis **46**
Sapindaceae 57
Savia erythroxyloides 16
Scaevola sericea 22, 37
Scolosanthus roulstonii **54**
Scrophulariaceae 58
Swietenia mahagoni 15

T

Tabernaemontana laurifolia **62**
Tecoma stans 36
Terminalia eriostachya **42**
        var. *margaretiae* **42**
Theophrastaceae 74
Thrinax radiata 15
Thymelaceae 75
Turneraceae 59
Turnera triglandulosa **59**
Turnera ulmifolia 59

V

Verbenaceae 60
Verbesina caymanensis **36**

INDEX OF COMMON NAMES

Page numbers indexed in bold type are the main description pages for the endemic and near-endemic species.

Australian Pine 21, 23, 37

Banana Orchid **52**
Beach Naupaka 22, 23, 37
Black Candlewood 47
Black Mastic **42**
Black Mangrove 15
Broadleaf **37**
Bull Hoof 34
Bull Thatch 15
Burn Nose **75**
Buttonwood 15, 21, 58

Cabbage Tree 15
Cat Bush 59
Cayman Silverbush **43**
Christmas Blossom **63**
Corato 16, **33**, 54

Duppy Gourd **64**

Ghost Orchid **50**

Headache Bush 47

Ironwood 20, **49**

Lady Hair **67**
Logwood 21

Mahogany 15

Old George 9, **38**

Poison Tree 15

Red Birch 13, 15
Red Mangrove 13, 15
Royal Palm 15

Sea Grape 16, 17
Silver Thatch **53**
Slingshot 62

Tea Banker **35**, 61

Wash Wood **74**
White Mangrove 15
Wild Cocoplum 16
Wild Jasmine **62**
Wild Oil Nut **66**

THE SISTER ISLES

LITTLE CAYMAN

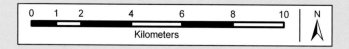